HILLARY ROSNER

ROAM

Wild Animals and the Race to Repair
Our Fractured World

patagonia

ROAM
Wild Animals and the Race to Repair Our Fractured World

Patagonia publishes a select list of titles on wilderness, wildlife, and outdoor sports that inspire and restore a connection to the natural world and encourage action to combat climate chaos.

© 2025 Hillary Rosner

Portions of this book were previously published in *bioGraphic* (Chapter 2) and *The New York Times* (Chapter 8).

Hardcover Edition

Published by Patagonia Works

Printed in Canada on 100 percent postconsumer recycled paper.

Editors: *Makenna Goodman, Sharon AvRutick*
Photo Editor: *Jane Sievert*
Art Director/Designer: *Christina Speed*
Project Manager: *Sonia Moore*
Photo Production: *Bernardo Salce*
Graphic Production: *Michaela Purcilly, Natausha Greenblott*
Creative Director: *Michael Leon*
Publisher: *Karla Olson*

Hardcover ISBN: 978-1-952338-31-1
E-Book ISBN: 978-1-952338-32-8
Library of Congress Control Number: 2025940439

Front Cover: A herd of elk at their winter grounds in the National Elk Refuge, outside Jackson, Wyoming. These elk migrate seasonally across the landscape, including in and out of Grand Teton National Park. *Florian Schulz*

Front Endpaper: Snow geese migrate thousands of miles round-trip each year. Long-distance migratory species need food and habitat across the length of their journey. *Konrad Wothe/NPL/Minden Pictures*

Title Page: A Marsican bear left a trace in a barbed-wire fence in Italy's Apennines. Only about sixty of these bears, a subspecies of brown bear, remain. Efforts to protect the species include taking down disused fencing. *Bruno D'Amicis*

Table of Contents: An Apennine wolf track on a forest trail in Abruzzo, Italy, in the Central Apennines. Roughly 2,400 wolves reside in Italy. *Bruno D'Amicis/Nature Picture Library*

DEDICATION

For Cooper—may you live on a planet teeming with wild life.

LAND ACKNOWLEDGMENT

Building mutual relationships with Indigenous peoples is part of our work to restore Earth, the home we all share. Patagonia's headquarters is located on the unceded homelands of the Chumash people in what is now known as Ventura, California. Because the people in this book are in locations around the planet, we acknowledge the many Indigenous communities who have stewarded the lands and waters of each of these places since time immemorial. We are also grateful for their continued leadership in the environmental and climate movement today.

RESOURCES

To learn more about and support the organizations doing the vital work discussed in this book, scan the QR code below. It will point you to their websites, as well as maps and additional resources.

6   AUTHOR'S NOTE

10   INTRODUCTION
The Adventures of
Lucky Pierre

26   CHAPTER 1
Paving (and Flooding)
Paradise: The Science
Behind Connectivity
Colorado's Front Range

48   CHAPTER 2
The Linear Path: How
Species Move (and Why
Understanding It Matters)
Montreal & South Carolina

82   CHAPTER 3
The Matrix: Creating
Climate Corridors So
Animals Can Move
Costa Rica's Osa Peninsula

118   CHAPTER 4
Entering the Agave Corridor:
Connecting Bats, Plants,
and People
Central & Northern Mexico

142   CHAPTER 5
A Wilder Europe: Bears,
Rewilding, and the Beauty
of Coexistence
Italy's Apennines & Trentino

176   CHAPTER 6
A Fence Runs Through
It: How Fences Make Us
Bad Neighbors
*Wyoming's Greater Yellowstone
Ecosystem*

198   CHAPTER 7
The Carnivore Among
Us: Coyotes and the
Urban Wilds
New York City & Long Island

224   CHAPTER 8
Corridors of Injustice: How
We Treat Each Other Is
How We Treat the World
Los Angeles & North Carolina

250   CHAPTER 9
Hidden Connections:
In the Everglades, Water
Is Everything
Florida

288   CHAPTER 10
The Time Is Now: Elephants
and the Gullies of Kenya
Kenya

318   CHAPTER 11
The Missing Link: Toads,
Caribou, and Us
Colorado & British Columbia

334   EPILOGUE
The Power of
Bearing Witness
Iceland's Kárahnjúkar Dam

342   ACKNOWLEDGMENTS

346   RESOURCES

347   INDEX

AUTHOR'S NOTE

In 2021, the year I began researching this book, Congress passed the Bipartisan Infrastructure Law, authorizing $1.2 trillion for infrastructure upgrades and repairs. That law included more than $28 billion for the Interior Department (which includes the US Forest Service, the USDA, the National Parks Service, the US Fish and Wildlife Service, and the Bureau of Land Management, among other agencies) for projects related to ecosystem restoration, scientific innovation, tribal investment, and much more. The law also included the Wildlife Crossings Pilot Program, with $350 million in federal funding to build safe passage for wild animals across dangerous highways.

As I write this, in May 2025, the world is radically altered. The current Interior secretary is planning to sell off large parcels of the nation's public lands—to be paved over and forever lost—while the Trump administration has slashed funding for conservation programs and the science that underpins them. To give just one example: In 2023, USAID, which Trump dismantled in one of his first actions, funded nearly $700 million in conservation and forestry programs around the world. All of that aid is now gone, projects stopped literally overnight. Years and decades of progress toward protecting and restoring the planet's imperiled ecosystems and species has been reversed. People who have devoted their lives to protecting our planet and its wealth of natural treasures have been fired, sidelined, demoralized.

In these pages, I chronicle people, projects, science, and species that have since lost their funding or are holding on by their fingernails. The habitat leases I write about in Chapter Six are in jeopardy, and related grants have already been cut; the main USDA program that funds conservation on farms and ranches

has been explicitly targeted for elimination. Everglades restoration activities I discuss in Chapter Nine are imperiled by job cuts at national parks and wildlife refuges as well as terminated lease agreements for the US Army Corps of Engineers. And don't even get me started on Chapter Eight, which is centered on equity and justice and the growing understanding that how we treat one another impacts the non-human species with whom we coexist.

Today, if you pull up virtually any Interior Department announcement from the Biden Administration—for example, one from September 2024 touting $92 million in funds for river and aquatic habitat restoration—you will see a giant banner across the page informing you that the content is "ARCHIVED" and that links no longer work. Click on a link from a sentence explaining this "major investment in the conservation and stewardship of America's public lands," and you'll get a 403 error message.

But funding conservation is not an error. Protecting the natural world is perhaps the most essential thing we must do as humans. "The greatest lie that humans ever told," wrote the British author Katherine Rundell, "is that the Earth is ours, and at our disposal."

What if we learned to regard other animals and the way they live with curiosity and empathy? What if we began to see the world through their eyes? What if we learned to coexist with the marvelous, majestic, magnificent creatures with whom we share this planet?

In this uncertain and disconcerting time, there is one thing about which I remain certain: Curiosity breeds hope.

I hope, by the time you read this, there has been good news for the planet. But I also hope this book sparks your own curiosity, and inspires you to help protect our natural world, in any way you can.

May 2025

INTRODUCTION
The Adventures of Lucky Pierre

At the tail end of the twentieth century, when I was living in the engineered asphalt ecosystem of Manhattan, a coyote made its way into Central Park. It was April Fool's Day, of all things, and the incident had the vague whiff of prank. Helicopters searched the ground from above, while a throng of cops and animal welfare workers chased the coyote across the park's landmarks, from the skating rink to the Great Lawn to a nature sanctuary and finally a World War I statue, where the bedraggled animal collapsed, having been hit by a tranquilizer dart as he tried to dodge his pursuers by swimming across a pond.

The coyote, whom authorities at first nicknamed Lucky Pierre (after the Pierre Hotel on nearby Fifth Avenue), was ultimately renamed Otis and sent to live in a half-acre enclosure in the Queens Zoo, where he lived in a pack of renegade coyotes captured in various places unsanctioned by humans. At the time, park officials noted that coyotes had become "an increasing presence in New York City, particularly in the Bronx," but they weren't certain how this one had arrived in the center of Manhattan. They speculated that he had trotted over the Henry Hudson Bridge, which connects Manhattan to the Bronx, but how he navigated the concrete canyons of the Upper West Side was a mystery.

I grew up in those same concrete canyons and had returned after college to work as a journalist, and I was transfixed by the idea that there might be ribbons of green spaces through

Previous Spread: A bird's-eye view of Manhattan. Otis the coyote's escapades in Central Park in the spring of 1999 first got Hillary thinking about how wild animals move through human landscapes. *Predrag Vuckovic/Getty Images*

which a reasonably large mammal could travel in such a heavily urban area—like a highway, but for animals. I had spent a good chunk of my childhood traipsing around Central Park, and now I looked at the place in a new way. I was fascinated that this manicured, purpose-built, fake-natural place could host wild creatures beyond rats and squirrels and the odd pond turtle. Otis was my first intellectual encounter with the notion of wildlife corridors.

As a kid, I loved when my family would take excursions to natural places outside the city, but I knew very little about environmental issues and learned almost nothing about ecology or the workings of the natural world in school. I couldn't see how the science we were taught in the classroom connected to my life—so for the most part, I stayed away from science. In high school, I opted for an extra English class in lieu of chemistry. In college, I tried to get over my science aversion by enrolling in "genes for jocks," only to earn a C. But around the same time as Otis's unfortunate adventure, I had started paying attention to reports of rainforest destruction—in Brazil, in Indonesia, in central Africa.

Environmental groups were running media campaigns that connected the dots between the things we Americans purchased (two-by-fours, soybeans, tissues) and the ancient trees and beautiful creatures that were vanishing, often on the other side of the planet. The closest I'd ever been to a rainforest was the Bronx Zoo's monkey house, but increasingly, all I wanted was to study how rainforests worked.

For a while, I would sit at my fluorescent-lit desk at *The Village Voice*, beneath an HVAC duct delivering its stale air, and pore over the course catalogs of forestry schools, dreaming. A bit later, when I briefly worked as an editor at a media website called Inside.com, I lamented the irony of working for Inside when I wanted to be outside. I wrote a tiny piece for *New York* magazine on how the scramble for tantalum—a mineral mined in Congo

for use in cell phones and other electronics—was endangering fragile populations of gorillas and other wildlife. It was maybe 400 words, and I never left my desk to report it. But I was hooked.

Within a couple of years, I had left New York for Boulder, Colorado, and enrolled in a master's program in environmental studies. I took a slew of classes about climate change and all its looming impacts. I audited extra courses to ingest the greatest possible amount of information. I learned about biogeochemical cycles and food webs, about wildlife management and natural resource law, about renewable energy and B Corps and payments for ecosystem services. The subject matter was often grim—but the potential solutions to environmental ills were plentiful and on the horizon. Time was of the essence, but surely we humans were going to set things right. There was just so much scientific knowledge at hand, so much technology available, so much enthusiasm to ward off disaster.

I kept writing, focusing my journalism on environmental issues. I eagerly tromped off to rainforests and deserts and tundra, in lands that had once seemed to me so distant they might as well have been imaginary—Ethiopia, Indonesia, Nicaragua, the remote reaches of the American West. I wrote scores of stories about climate change and its impacts and solutions: about an insect decimating pine forests where trees were weak from drought and the bugs were thriving in warmer temperatures, about people working to breed new crops that could thrive on a changing planet, about efforts to harness bacteria for fuel.

But the stories that really grabbed me were always, in some way, about habitat fragmentation. I was drawn to the tales of creatures whose existence was in peril due to a shrinking ability to roam—whether that roaming was on the ground, underwater, or through the air. I wrote about a native Colorado River fish, the razorback sucker, trapped by a dam and unable to fend off invasive predators in the reservoir where it was now

forever confined. I wrote about wildlife roaming a narrow riverside corridor in Borneo because the unchecked expansion of palm oil plantations had marooned them on the banks. I wrote about saline lakes of the Great Basin, critical stopping points for migrating waterbirds, mysteriously drying up. I was struck, again and again, by the dramatic scope of land-use change—the fundamentally important matter of how we are slicing up and tearing down and paving over the planet.

Like many people involved in conservation issues at the time, I naively thought that knowledge alone could set things right. Surely now that the scope of deforestation was clear, I reasoned, we would stop bulldozing and burning the jungle. Surely we would save the remaining tall-grass prairie, the free-flowing rivers, the boreal forest, the coral reefs. But the rate of decline only accelerated. In the first decade and a half of the twenty-first century, humans obliterated or sliced apart more than 10 percent of the globe's largest remaining swaths of forest.

— – – —

Over the past decade, climate change has come to dominate media coverage of the environment, and much of the environmental agenda. It's such an all-consuming, essential, and existential problem that it overshadows many other issues. But land-use change—the way we alter the Earth's surface—is also critically important, and especially because it interacts with climate change in so many ways. Cutting down forests releases carbon dioxide into the atmosphere; covering the ground in asphalt makes it absorb more heat, exacerbating the impacts of already-rising temperatures; destroying wetlands leaves the ground unable to absorb water when extreme rainfall arrives as a result of weather patterns changed by greenhouse gas emissions.

Against these backdrops, as the climate shifts, species all over the globe—plants and animals alike—will need to find new habitat that meets their needs. It is already happening. Some

will have to migrate increasingly longer distances to find suitable homes. And in between their current home in one protected area and a feasible future home in another, there might be a city. Or a barbed-wire fence. Or an unscalable international border wall. Or a dearth of animals and birds to help spread the seeds of future trees and plants they'll need to survive.

Almost invariably, there will be a road.

Now consider that paved roads, the quintessential icon of human connectivity, serve this function for no other species but us. When you start to really internalize that fact, it's impossible not to see the world in a whole new light. Think about it: For most other species on the planet, the roads that connect us serve the exact *opposite* function. They are a barrier, a deterrent, a danger, a hard stop. For many species around the world, the other side of a busy highway may as well be another planet.

It's true that humans have always altered the landscapes around us, for better and worse. Over millennia, for instance, people used fire to manage vegetation, cultivating the land for our needs. We also arguably wiped out an entire planet's worth of large mammals that once cavorted across continents. (A chilling climate helped, but we were at the very least a major driver.) More recently, colonialism and its singular focus on resource extraction—its view of the Earth and everything on it as nothing but loot—ripped minerals, timber, wildlife, and people from their origins and set in motion a centuries-long trend of dismantling ecosystems and Indigenous cultures around the world.

But over the past few decades, the pace of our changes has spun out of control. Academics call it the Great Acceleration,

---

Otis sizes up his habitat at the Queens Zoo. He most likely migrated south from leafy Westchester County and into the Bronx, moving through green spaces where possible, before ending up in Central Park, where his presence sparked a chase and a media frenzy. *Bill Turnbull/NY Daily News Archive/Getty Images*

the period since the 1950s that has seen a huge increase in the rate of land conversion, deforestation, and biodiversity loss, alongside similar spikes in population, energy use, water use, fertilizer use, dam building, overfishing, urbanization, and, of course, greenhouse gas emissions. Humans have literally changed the face of the Earth, on an overwhelming scale, through industrial scale farming, thoughtless sprawl, highways, mining, drilling, and erecting millions of miles of fences. We've done this, too, through our relationships with one another, by tying opportunity and access to identity—behavior that influences the built environment and in turn impacts other species as well.

And we've done it by refusing to recognize the interconnected nature of the planet's systems, and by making short-term and short-sighted decisions about building and development, without considering the collective impact of millions of people clearing and paving and fencing their own slices of the planet. We have prioritized humans over other beings, despite knowing that our fate relies on the interconnectedness of us all. We are also in the process of reshaping the world through climate change itself. Melting glaciers, rising seas, burning forests, drying-up lakes—all of these have an extreme impact on the way animals move around the planet.

I don't want to irreversibly depress you before we even get to Chapter One—I promise there will be hope!—but it's crucial that we recognize the scope of the problem. Increasingly, animals have nowhere left to go. About 40 percent of the planet's land has been converted to agricultural production. Cities, which cover about 3 percent of the planet's land, are expanding: Over the past three decades, urban areas have increased globally by 80 percent. One recent study projected that urban land area could increase by another 100 to 600 percent by the end of the century. Between 1992 and 2000, we converted more

than 73,000 square miles of the planet from "natural habitat" to urban areas, according to another study, which predicted that the period from 2000 to 2030 would see an additional 112,000 square miles converted. Less than 15 percent of land around the globe is protected from development—and fewer than half of the chunks in that 15 percent are connected to one another. In the US, only 41 percent of existing "natural" areas are sufficiently connected so that species can access the environmental conditions they need. And the lack of connectedness applies not just to land, but to water.

Ninety-three percent of the total volume of water that moves through the Earth's rivers is altered from its natural flow by dams—which chop up roughly 60 percent of the world's rivers. And an additional 3,700 new hydropower dams are currently planned or under construction worldwide. Of all the world's rivers longer than 1,000 kilometers, fewer than a quarter still flow to the ocean without interruption. And that's just the longest rivers.

If that wasn't enough, a large study published in the journal *Science* (coauthored by 115 scientists representing 24 countries) revealed that across the planet, in landscapes dominated by humans, animals of all kinds are roaming less. That finding has serious consequences not just for the fifty-seven species the paper tracked, and the many more it didn't, but for the ecosystems they inhabit—and for our own welfare.

The result of all this, for non-human species, is clear: As we carve up the planet, we close off the available pathways for other species to move. That's bad news for biodiversity, but it's also bad news for humans. Because when we cut down forests or bisect them with roads or farms, we create new ways for pathogens to spill over to people, aiding the spread of nasty things like Lyme disease, Ebola, malaria, and Covid. Ecosystems missing key species of plants or animals unravel, leaving them

defenseless against pests, more likely to burn in catastrophic fires, and vulnerable to all sorts of other misfortunes. Despite humans' great ingenuity, the reality is that we depend on functioning ecosystems for our most essential things: clean water, clean air, healthy soil, pollination, flood control, and on and on. We like to think we're above it all, but humans are deeply reliant on nature. We are also deeply connected to the millions of other species on Earth. My favorite phrase for this is "umbilical connectivity," a term I first saw in a United Nations report (a literary genre not generally known for its colorful language). Humans, the report said, are "disconnected" from our relationships with other living beings. We have forgotten our interdependence with the natural world. And this disregard has drastic consequences.

Connectivity is the subject of this book. It is also, unfortunately, one of those awkward words that science is filled with: too many syllables, too abstract to mean much, at face value, to the average person. But there's really no other word that quite fits. Connectivity means "the state or extent of being connected or interconnected." And that's the heart of the matter. Where will we be in another twenty years if we don't start knitting the natural world back together? Is it too late to reconstruct a world where animals can roam freely? How can we reconnect landscapes so that wild species and natural systems have room to adapt and change? And can we do it in a way that recognizes the long hands of racism and colonialism in creating today's ecological crises, and helps heal rather than deepen those scars?

Roam is my attempt to answer these questions. The answers are particularly urgent now—because, at a time when "connectivity" has become a buzzword to describe all the great ways humans are linked to one another through digital means, the truth is our physical planet, and our own relationship to it, has

never been more chopped up, detached, and disconnected. To attempt to reconnect is an act of hope.

This book is about how we can forge new links between landscapes that have become isolated pieces—what scientists sometimes call linear connectivity. It's also about how we can mend ecosystems back together, so that the processes still work and the systems can evolve as they need to—functional connectivity. Together, these two ideas make up the concept of ecological connectivity, which the International Union for Conservation of Nature (IUCN) describes as "the unimpeded movement of species and the flow of natural processes that sustain life on Earth." I'm also concerned with emotional and cognitive connectivity: how we can build a world in which humans recognize their interconnectedness with the rest of the planet and view other species with empathy and compassion.

Roam explores the vital importance of all these kinds of connectivity in the natural world. It's about the ways non-human species move around the planet, and how our roads and dams and fences and farms and mines and subdivisions have a cascading impact on virtually every animal—and plant—in the world. I hope it leads you to wonder about where wild animals "belong," or rather, where we are willing to tolerate them, where we won't, and how we can help ensure there is enough usable space for all of us. And I hope it causes you to stop and think about the decisions we make—both directly and indirectly, through lack of smart planning—about the landscapes we inhabit. The good news is that the science that underpins these decisions is advancing by the day, even birthing entirely new and fascinating fields like movement ecology (the science of how animals move across a landscape) and road ecology (the science of how roads alter ecosystems), while also embracing ecological knowledge passed down through generations of people living in a landscape.

I set out to look for signs of hope amid all the bad news, and to determine whether, and how, we might still be able to fashion something strong and whole, if not seamless, from the tatters. The journey took me from my own home on the Front Range of Colorado to other parts of the region and the country, to Central America, Europe, and East Africa. I interviewed more than 200 people and drew on interviews with several dozen others from stories I reported in the past that shaped my thinking. I was reminded, constantly, of how many amazing humans have dedicated their lives to protecting nature and its wonders, people who put in long, uncomfortable hours in the field, over and over again, because that's what it takes for us all to survive.

Often when reporting on science, I'm reminded of a picture book starring Curious George, where the man with the yellow hat takes George along with some museum scientists to dig for dinosaur bones. George watches as a scientist digs up dirt and sifts it with a sieve, looking for bones. She finds none. She digs again and sifts some more—nothing—and continues like this, digging and sifting and digging and sifting, for what seems to Curious George an eternity. I know that feeling. "George yawned," reads the story. "So far digging for dinosaurs was not as exciting as he'd expected."

Spoiler alert: Curious George accidentally topples a wheelbarrow full of dirt, runs away in embarrassment, and sets off a rockslide, revealing an entire dinosaur skeleton. While actual science might not unfold in quite this way, the book makes an important point: Although the scientific process can be tedious, scientific *results* can be not just exciting but groundbreaking (sometimes literally). They can finally offer proof for something

A radio-collared female coyote crosses a railroad bridge in West Chicago. These bridges often serve as links between otherwise disconnected green spaces in urban areas, allowing coyotes to navigate human-engineered environments. *Corey Arnold*

long suspected. They might overturn assumptions, compelling us to look at things in a new way. They can illuminate something amazing happening in nature, previously unknown to most of us, which helps bring our world into sharper focus. But the legwork involved in producing those results can involve long hours trudging through poison oak, fighting off mosquitos, sweltering under an unrelenting sun. It can lead to injuries or intestinal parasites. And it can require Zen-like patience, as you wait years for natural processes to yield answers. All of which involves a baseline amount of hope, and sometimes deep reserves of it as well.

Hope can be hard to sustain at times—something I've learned firsthand over more than two decades of reporting on the environment. But there *are* still reasons to hope—perhaps the strongest being that we cannot afford not to. To hope is not passive. It is an act of faith, an act of resistance. All around the world, there are people at all levels of power who are working to make the planet healthier and more resilient for all the organisms that call it home. There are efforts to protect some of the planet's largest and most ecologically intact areas from industrial development; to link up fragments of crucial wildlife habitat into vital corridors; to turn abandoned or messed-up landscapes back into thriving ecosystems; and so much more. There are people who think in connections, rather than fragmentations. Hoping is itself an act of connectivity.

— – – —

Okay, sure, Otis the coyote's Central Park bacchanal ended behind bars, which would perhaps not have been his first choice. But his journey, and that of other coyotes who followed later (and there were many more!), hints at the power of linking up green spaces. Because nature *wants* to survive. Given half a chance, nature can bounce back. But we need to make sure we give it that half a chance, at the very least.

The decisions we make and the ways that we live don't always reflect it, but most people want to see nature thrive. Most people understand subconsciously that when the Earth thrives, so can we. When we are depressed or struggling or need to recharge, we often seek out natural places. If you ask anyone privileged enough to have traveled on vacation about their favorite trips, they will frequently tell you about wildlife and natural phenomena: seeing elephants in the wild, whales breaching, a rare orchid blooming, a forest of giant trees. Who, upon learning they have a terminal disease, wants to take the trip of a lifetime to a shopping mall? Connection to the natural world is deep in our DNA. And for me, keeping faith in connection means understanding how we can make better land-use decisions, and studying the innovators who are using every possible tool to knit nature back together, be it technology, theory, talk—or even ropes and screws.

— – – —

A quick note about the chapters that follow: Connectivity is a huge, unwieldy topic with thousands of scientific studies devoted to it. Scores of conservation organizations of every size are focused on restoring connectivity; it's a topic that plays out not just on land but in rivers and seas across the planet. It is not my goal to write an "everything there is to know about connectivity" book. I have chosen stories that reflect what I feel are some of the most important and compelling issues of the present moment. As an organizing principle, I have focused almost entirely on terrestrial matters, consciously though lamentably leaving out nearly three-quarters of the planet's surface. I have traveled to the places that fascinated me and brought back stories of what I witnessed there, recognizing that I occupy a particular position as a White, Western journalist.

I hope what I found connects you to the natural world, which needs your help.

CHAPTER 1   *Colorado's Front Range*

Paving (and Flooding) Paradise: The Science Behind Connectivity

On an unseasonably warm October afternoon in 2021, I stood on a hilltop in northern Colorado and stared into a giant pit. Down on the valley floor below, scores of trucks—dump trucks, backhoes, diggers with enormous claws—snaked along on dirt tracks, moving earth and rock and any remaining sign of vegetation. From above, the trucks looked almost like toys, resembling a scene my son might have once assembled on the living room floor. But this was very real and very much life-size. The pit spread across 740 acres of land, an area slightly smaller than New York's Central Park.

I had come to this lookout spot outside the town of Loveland—a name that suddenly seemed absurd—along with a handful of journalists and a representative from Northern

---

Previous Spread: Colorado's Front Range—the north-south axis where the high plains meet the mountains—has exploded with growth over the last several decades, crowding out wildlife. Here, a subdivision eats up the prairie in the town of Superior, near Boulder. *Rick Wilking/Reuters*

Water, a public utility that supplies water to more than a million people. We'd come to see the construction of a new dam and reservoir. In a few years, Chimney Hollow Reservoir would be able to store 90,000 acre-feet of water to serve the area's mushrooming towns. An acre-foot is a measurement that covers one acre with one foot of water, and to visualize that, water people sometimes talk in terms of football fields. Chimney Hollow will hold enough water to flood more than 68,000 football fields with liquid to your shin. Or, if you want to think in more tangible containers, that's about 29 billion gallon jugs full of water.

Water in the western United States is an increasingly scarce resource, and the population is growing. The population of Fort Collins, north of Loveland, more than tripled between 1960 and 1990, to about 88,000 residents, and then doubled again over the next thirty years, to around 170,000 in 2022. In 2024, the median price of a house there was $555,000, up from about $420,000 just five years earlier. A bit further out on Colorado's eastern plains, formerly rural communities like Frederick and Dacono are now bustling suburban towns filled with residents flocking from around the country. Over the last decade, Frederick's population grew by 67 percent and Dacono's by more than 50 percent.

All those people need water, and Northern Water's job is to supply it. Chimney Hollow is part of a sprawling system of tunnels and pipes and reservoirs that transport water from the Colorado River, on the western side of the Rockies, to the towns and cities of the Front Range, on the eastern side. There, the water must be stored. The whole system, known as the Colorado–Big Thompson Project, was begun in 1938 and took nearly twenty years to complete. Chimney Hollow is among a series of new improvements meant to help shore up the water supply for the twenty-first century. It's intended as something

like a giant bucket to store extra water during wet years for use during dry ones.

I first visited Chimney Hollow in 2015 while writing about the new era of dam construction in the West. Global warming was already causing prolonged droughts and shifting the timing of snowmelt, which becomes the water that flows into creeks and rivers. Governments were starting to worry about dwindling water supplies. At the time, I'd asked a longtime employee of Northern Water if by building huge new infrastructure to store water we were tacitly telling people that the population growth in these Western towns was sustainable. "Some people think if we don't build those projects, people just won't come," he told me. "I wish that were the case. But it's not gonna happen. People are going to keep moving here, because it's a great place to live." Having moved to Colorado myself in 2002 from New York City, I couldn't really argue.

Since 2002, I'd personally watched Colorado's Front Range— the north-south axis where the high plains meet the mountains— explode with growth. Each time I drove around the area, I saw new construction, new bare dirt, new bulldozers. Land that I'd long assumed was protected from development turned out not to be—it just hadn't been developed yet. Sprawling meadows became sprawling subdivisions. When I moved to the Front Range, a newly constructed toll road to Denver International Airport was an isolated river of asphalt across empty rolling plains. You could drive for miles across farmland, ranchland, shortgrass prairie where hawks stalked their prey. Today, driving to the airport, it's nearly impossible to conjure that image. Those farms and meadows have been replaced by houses,

---

In 2015, when Hillary first visited the site that would become Chimney Hollow Reservoir, near Loveland, Colorado, she found it difficult to imagine that this place would soon disappear. *Hillary Rosner*

shopping centers, hotels and hospitals, acres and miles of soil covered over with asphalt, no longer providing forage and nesting grounds and safe harbor for the animals that once lived there. You don't have to be a scientist to look at those vistas and realize we have made it much, much harder to be a wild animal.

But my perspective on development is based on a particular moment in time. I lamented the transformation this new road had set in motion, but I didn't complain about the highways that *already* stretched across human-made landscapes. We rarely look around at our own surroundings and imagine them as they once looked, before colonists stole land from Indigenous people and then cleared trees and began building dams and ditches and dirt roads, or before our more recent predecessors built highways and bridges and power plants and other modern monuments to automobiles and consumption. My own house sits on what was once a prairie, home to buffalo and grizzly bears and the people who coexisted with them. Today, it is home to a high concentration of Rivians and multimillion-dollar homes with manicured gardens. Sure, black bears and bobcats wander through, evoking a sense of the wild. But the place is radically transformed.

And although some neighbors recently tried to stop the subdividing of an acre lot lush with hundred-year-old trees—where bears often foraged and migrating birds stopped to rest—I've never heard anyone, myself included, complain about the existence of the neighborhood itself, or the cemetery or government labs that abut its southern edge, or the city that spreads out all around it. The people who ultimately move into the new houses built on the former wooded acre will marvel at the few tall trees still standing in their yards and know nothing of the small forest that stood there just a couple years earlier. It's the shifting baseline phenomenon: We take our own moment for granted as a starting point.

This has important implications when it comes to conservation, because the overarching trend for the past several hundred years has been the carving up of natural ecosystems (and the forced exile of humans who lived in those places respectfully for millennia). The more we carve things up, the harder it is to put the pieces back together, particularly if your baseline is already something less than whole. It's just as true in my own backyard as in Amazonian rainforests, the African savanna, and everywhere in between. Think about Humpty Dumpty sitting on the wall: If he simply broke in half, it would be a hell of a lot easier to put him back together than if each of those halves shattered, and those shards broke into still smaller pieces, until no one alive could recall anymore how to glue it all back together or even what the slivers had amounted to.

— – – —

I wasn't around when the older infrastructure of the Big Thompson project was built. But I did watch a new reservoir being gouged into a gentle valley, and I couldn't shake the image. While reporting that water story, I visited Chimney Hollow before any bulldozers arrived. I walked the valley floor and tried to etch it in my mind by noting its smells, its sounds, the motion of the wind. The place was a vast meadow awash in the mustardy yellows of late summer in Colorado. Visitors the previous week had reported seeing a bear and the tracks of a mountain lion. Ponderosa pines dotted the landscape, birds hopped and twittered in the grass and shrubs, and at the lowest elevation an old stream bed was lined with cottonwoods. I was struck by the valley's beauty: the ochre, the grass in the breeze, the birdsong. It was hard to fathom that this place would soon disappear.

At the time, I'd been consumed by what the reservoir might mean for the wildlife that called Chimney Hollow home. What would happen to those bears and mountain lions, the snakes and frogs and small mammals that dug tunnels underground,

when the whole place was underwater? It was tough to fathom what the transformation might entail. I drove home that day feeling what I described to my husband as a kind of heartbreak. I knew there were no easy answers, that opposing the reservoir amounted to hypocrisy at some level. I myself grew thirsty garden tomatoes, took longer showers than necessary, even watered my lawn during a drought. But standing there, I was certain of something: I wished there was some way to spare this beautiful little place.

But there was not. Standing on the hillside again in 2021, I pulled out my phone and found a photo I'd taken on that earlier visit. I tried to reconcile the "before" picture with the industrial dystopia that lay below me. It was the same geographic place, but it was deeply altered—not unlike the broader world that also seemed in that year to be unsettlingly altered, by a global pandemic, by the piling up of climate-related disasters and wars, by the proliferation of conspiracy theories. To see nature being upended in such a way, ripped apart and completely transformed, gave me a strange sense of panic. It reminded me of a feeling that sometimes afflicts me while traveling, of waking up in a hotel room and momentarily having no idea where I am or where the door is. It was beyond disorienting to see such a magical place transform into its industrialized doppelganger.

As before, I couldn't help but think of the wildlife. I tried to imagine what this would look like from the point of view of an elk, or a bobcat, or a small mammal like a mouse or vole. Would an elk experience this same kind of disorientation if she had crossed this meadow previously and came back to find it either turned into a wasteland of belching machinery or mutated into

Chimney Hollow Reservoir under construction in 2021.
It was hard to reconcile this giant industrial pit with the
beautiful meadow, teeming with life, that was here before.
Hillary Rosner

a lake? Needing to cross, to continue a journey to the east or the west, would a bear walk three miles to the south in search of a way around? Where would the vole go when his tunnel home began shaking and collapsing and finally filled with water?

The answer to the age-old riddle, "Why did the chicken cross the road?" is, of course, to get to the other side. But what if the chicken—or the sage grouse, or the lesser prairie chicken, or any wild creature accustomed to roaming unhindered— encountered a parade of semi-trucks barreling by, threatening to squash it? What if the road was suddenly a 29-billion-gallon lake? These might seem like naive, trite questions—the stuff of jokes, even. Questions posed by a clueless, head-in-the-sand tree hugger who worries about furry animals when the people of the West need water. But we can't continue to think in binaries. An empathy for other species is essential as we work to protect the ecosystems we all share. It's not animals versus humans. We are all on the same side.

From childhood, we're taught the phrase "bird's-eye view." But how often do we really try to see the world from the perspective of a bird? Or a bear, a frog, a sloth, a moth, an elephant? Do we ever consider how it feels to wake up as a jaguar in a Central American forest and spend your day searching for food or a mate? What happens when you, the chicken—or the frog, or the jaguar—come to the highway, or the fence, or the subdivision? What happens when the forest you live in and move through succumbs, piece by piece, to logging trucks, or, all at once, catches on fire?

These sorts of questions are just one way of thinking about the idea of connectivity. To survive, wild animals must move around in an increasingly human-modified world. That movement can involve a whole range of types and needs. There's daily travel, which is simply the ways in which animals move through a landscape to meet their everyday requirements. Think of a backyard

squirrel foraging and storing food, finding water, building a nest, raising a family. There's migration, which is the seasonal movement of groups of animals between different areas. Think of a wildebeest moving 1,000 miles across East Africa, or a monarch butterfly migrating roughly 3,000 miles from Canada to Mexico (a complete migration cycle that actually involves four to five generations). Think of Arctic terns, with the world's longest migration—somewhere between 25,000 and 50,000 miles—traveling between Antarctica and Greenland. What do they each need to do to get to their "other side"?

There's also a concept called "dispersal," when an individual animal leaves its family or pack and sets out to find a new home territory. Dispersal helps ensure healthy gene pools. And there's the strange and slippery concept of "future movement," a response to climate change in which animals will need to move in the future as their existing habitat no longer provides what they need. That future has already begun. To ensure animals can move where they need to, conservation will increasingly require us to predict their pathways and end points, and to make certain those places don't all wind up as reservoirs, roads, or parking lots.

— – – —

Back in the 1960s, biologists E. O. Wilson and Robert MacArthur developed the theory of island biogeography, a way to explain the distribution of species around the world. The theory proposed that large islands would house more species than smaller islands, and that extinction rates would be higher on smaller islands because there would be fewer members of each species, making it more likely that they might all be wiped out. The theory also proposed that islands that were closer to a mainland would have more species and lower rates of extinction, because those islands would be easier for animals from the mainland to colonize. While island biogeography theory was about real,

actual islands, it turned out to have important implications for fragments of habitat that are *metaphorical* islands. It explained why splintered habitats tended to have higher extinction rates than those left intact. When you slice up an ecosystem, leaving behind disconnected fragments of nature, those fragments function like islands, leaving populations of animals and plants there marooned and vulnerable to everything from disease to localized environmental disasters to inbreeding.

Not long after Wilson and MacArthur published their theory, a biologist named Richard Levins became interested in melding seemingly disparate fields to find a unifying theory involving ecology and evolutionary biology and formulated the idea of "metapopulations." A metapopulation is a system of local populations spread across a landscape, and Levins's theory dealt with the rise and fall of those populations—how they grew or went extinct, how the changing dynamics of those patterns influenced a species' ability to exist across a region. Metapopulation theory, as McGill University ecologist Andrew Gonzalez put it to me, was better adapted than island biogeography to real-world conservation scenarios, "where habitat is heavily fragmented" and made up of a patchwork of disconnected landscapes that are home to "extinction-prone populations."

Over the next few decades, ecologists latched on to metapopulation theory and applied it to conservation science through the lens of connectivity. A Finnish ecologist named Ilkka Hanski spent several decades combining field data and statistical models to explain what could happen to metapopulations in "highly fragmented landscapes." He introduced the idea of "metapopulation capacity," which measures a landscape's ability to sustain healthy metapopulations. Scientists developed other concepts helpful to connectivity conservation, like "extinction debt," the disturbing idea that habitat destruction in the present will ultimately lead to extinction in the future, but that there is a

time lag during which a species can hang on despite its fate being sealed.

In the late 1990s, the metapopulation idea evolved into the concept of "metacommunities." A metacommunity approach views ecosystems as networks in which metapopulations of one species interact with metapopulations of other species; these between-species interactions can contribute to the risk of extinction. The species that are interacting could be a predator and its prey—say, wolves and elk. Or it could be animals that compete for food, such as squirrels and chipmunks. Or it could be the multiple species that interact to spread a disease, such as the bacteria that cause Lyme disease, the ticks that carry the bacteria, and the deer that move the ticks around—and also the humans the bacteria infect. A network's "size, configuration, and connectivity," as Gonzalez put it, "defines the species and processes that persist there."

Over the decades since Wilson and MacArthur developed island biogeography theory, as the science has continued to evolve, research has repeatedly demonstrated the same thing: Fragmenting habitat into islands leads to huge losses of biodiversity. And as human population has soared across the world, the question of how to best protect habitat for other species has grown in prominence. Faced with an onslaught of industrialized development, the question becomes a choice: What is most important to save? In the 1970s, the question erupted into a scientific debate that became known as "SLOSS," which stands for "single large or several small." Was it better to protect a few giant chunks of land or many smaller patches that added up to the same area?

Thomas Lovejoy, the American biologist who coined the term "biological diversity" and brought global attention to the destruction of the Amazon's forests, set out to answer the SLOSS question in the late 1970s. (Lovejoy passed away in late

2021, in the same week as Wilson—a huge loss for conservation in a single week.) He took advantage of a new Brazilian law mandating that landowners who were slashing and burning the Amazon's rainforests to raise soybeans and cattle had to leave 50 percent of their property as forest. He managed to convince some landowners, and the government agency in charge, to configure that 50 percent into a research project. The experiment, which is still running more than four decades later, created a series of forest fragments of three different sizes—2.5, 25, and 250 acres—isolated in a sea of clear-cuts.

A story about Lovejoy's Amazon Biodiversity Center, written by the journalist and conservationist Amanda Paulson in 2018, quoted an ornithologist working there who said the smallest fragments were "imploding over time." Paulson described the strange sensation she felt inside one of the medium-size forest blocks. "As we walk into one 25-acre fragment, it still looks like a dense forest, but it's eerily quiet," she wrote. "No monkeys reside here. We hear some cicadas and the distinctive call of the screaming piha, but otherwise the forest seems largely devoid of life. Even the trees in these individual plots grow more slowly."

In reality, the larger plots of 250 acres were too small for some birds to survive in. Within fifteen years, half the bird species had vanished. Black spider monkeys disappeared from all the forest fragments, too. As Lovejoy put it in *Ever Green*, a book he penned with the economist and conservationist John W. Reid, these monkeys "move fast through large areas of forest eating fruit from widely spaced trees." That was no longer possible in any of the fragments. White-plumed antbirds also abandoned even the larger patches. These birds exist in a symbiotic relationship with army ants, following their invading columns and dining on the bugs that try to flee. Each ant colony only went on a rampage every few weeks, explained Lovejoy and Reid, so the antbirds needed to link up with multiple battalions

and follow their invasions in turn. "The 250-acre fragments were at least three times too small for the birds. No antbirds means no antbird droppings, which deprives shimmering blue-and-black butterflies their sustenance. They left, too."

The results from the Amazon project have been unequivocal. As Lovejoy summed it up in *Ever Green,* which was published posthumously, "Large intact areas are very important, the larger the better." It is also true that there are some species for which a greater number of smaller patches may be more import-ant, and that there are places where small patches of habitat are all that remain, making their conservation vital. One group of scientists put it best, writing, "the best plan is to set aside as many reserves as possible, and that they should be as large and connected as possible." Today, Lovejoy's research site covers 620 square miles and has generated hundreds upon hundreds of scientific papers. And while the findings have sounded the alarm with a host of bad news about fragmentation, they have also illuminated the path forward.

Over the years, more and more research has piled up to show that there is a direct relationship between the size of a reserve and the long-term stability of species there, and that areas at the edge of forests are most susceptible to ecological damage. The more fragments you have, the more edges there will be. In a paper published in 2015, a group of scientists—Lovejoy among them—looked at the effects of habitat fragmentation on ecosys-tems around the globe. They concluded that 70 percent of the

---

Next Spread: In the Brazilian state of Pará, south of Santarém, the Amazon rainforest has been razed for soybean production. Deforestation rates have been declining recently, thanks in part to better enforcement of environmental laws. But researchers worry the Amazon could be reaching a tipping point, where so much of the forest is lost that the entire system will break down. The Amazon is home to a tenth of the planet's biodiversity and stores vast amounts of carbon.
Daniel Beltrá/Greenpeace

forest that remains is within a kilometer of the forest's edge—
the most vulnerable region. "These findings indicate an urgent
need for conservation and restoration measures to improve
landscape connectivity," they wrote.

In addition to Lovejoy's Amazon experiment, several other
long-term connectivity experiments around the world have
helped amass huge amounts of data on the importance of connectivity for a whole range of species of mammals, birds, reptiles,
insects, and plants. In a series of forty-eight enclosures in a small
town near the Pyrenees, in southwestern France, scientists have
spent a decade studying the impacts of fragmentation and corridors on small critters like butterflies, dragonflies, toads, and
lizards. In the Danum Valley in Malaysian Borneo, researchers
have scrutinized the effects of logging, clear-cutting, and conversion of forest to palm oil plantations. On the site of a former
nuclear weapons facility in western South Carolina, researchers have spent thirty years replanting savannas and studying the
progress, trying to avoid poison oak while meticulously documenting the impact of corridors on plants and insects.

All these experiments have resulted in more or less the same
finding: Chopping up ecosystems slashes their biodiversity and
alters the entire system's ability to function. The smaller and
more isolated the fragment, the greater the impacts—and those
impacts only get worse over time. But they have also shown
that keeping ecosystems intact, restoring them, and reconnecting the fragments does a world of good.

That idea has also been proven by on-the-ground projects
dedicated to connectivity conservation. The Yellowstone to

Wolves in Alaska's Denali National Park. Even a six-million-
acre park isn't large enough to support all the wolf packs that
rely on its resources—which is why it's crucial to protect and
link habitat on and off public lands and across massive areas.
Florian Schulz

Yukon Conservation Initiative, or Y2Y, one of the earliest projects formed around a goal of protecting large areas for mammals like grizzly bears and wolves to roam, spans a roughly 2,000-mile expanse from Yellowstone National Park up to the Arctic Circle in Canada's Yukon Territory. Begun in 1993 as an idea that Canadian conservationist Harvey Locke scribbled down at a campfire after spending a few weeks traveling on foot and horseback through British Columbia's Northern Rockies, Y2Y took a vision of continent-wide conservation first promoted by scientists in a group called The Wildlands Project and made it flesh. In Y2Y's first twenty-five years, protected areas across its vast area increased by 80 percent, grizzly bear populations more than quadrupled, and at least ten major highway overpasses and more than a hundred underpasses were constructed.

— – – —

Three years after my first disorienting visit to the reservoir overlook, in the fall of 2024, I visited Chimney Hollow again. I had been back once in the interim, in the spring of 2023, when construction was moving along. At that time, the 3,700-foot-long dam's asphalt core was in place, and there were on-site asphalt and concrete plants as well as a quarry capable of supplying 63,000 tons of rock—4,500 dump trucks' worth—per day. The overlook had been outfitted with official signage about the project, which touted its environmental commitments. These amounted to "improved streamflow and aquatic habitat" and "providing West Slope water supplies"—water for people whose water we Eastern Slopers were siphoning. There was no language at all about land-use change or habitat loss. Now, another year later, on a dazzling but chilly day, I was disoriented once more. As we drove up the road toward the construction area, I was suddenly jarred by the sight of a giant wall. Sitting in the passenger seat, I had been anticipating the turnoff toward the reservoir—so where did this wall come from? Had I

taken my eyes off the road for a moment and mistaken our location? Then it hit me: The giant wall was the dam.

It rose 300 feet from the valley bottom now, like a thirty-story building jammed in between two hillsides. When we arrived at the overlook, gazing down at it from above, it was hard to reconcile its height. It just looked like a big flat road atop a berm. But from the ground it was shocking, a whole valley walled off at one end. I was overcome with solastalgia—a term coined in the early twenty-first century by an Australian environmental philosopher named Glenn Albrecht. Albrecht described solastalgia as "the distress that is produced by environmental change impacting on people while they are directly connected to their home environment," which he said was exacerbated by "a sense of powerlessness or lack of control over the unfolding change process." It was just one reservoir. But it was a stark illustration of the radical re-forming we do to the Earth, and why thinking about connectivity is so important.

The Linear Path: How Species Move (and Why Understanding It Matters)

One chilly evening during the height of the pandemic, I was snuggled up in bed with my son, who was seven years old at the time. He was reading a graphic novel involving aliens and kids in outer space. I was reading an academic book about corridors and connectivity. He looked up and noticed the title in my hands. "Mama, what's a corridor?" he asked. I thought for a moment. How to explain this to a child?

"Picture a forest," I told him. "And imagine that before people built a city, with all the schools and stores and parking lots, that whole area was a forest, where animals just roamed around doing their animal things. And then one day someone came and cut down a lot of trees and built a road across the forest. And now the animals on one side were separated from their friends on the other side. If they tried to cross the road, they might get

---

Previous Spread: Pronghorn migrate across a highway that bisects their route in western Wyoming. Roads may connect humans, but they serve as barriers to the rest of life on Earth. *Joe Riis*

hit by a car." A corridor, I told him, might simply be a safe way for the animals to get across the road.

"Like one of those bridges over a highway that animals walk on?" he asked. "Exactly like that," I said. I asked him where he learned about those. "In a book in school," he said matter-of-factly. He seemed to mull it over for a moment longer, and then went back to the aliens. I was thrilled that my son was learning about wildlife corridors.

A corridor is the most basic, tangible way to think about connectivity. It's a pathway to get from one area of habitat to another. It could be a bridge or tunnel across a dangerous road, or a ribbon of green through a heavily developed area, or a system of linked natural areas connecting protected reserves. Corridors are often the easiest form of connectivity to visualize, especially when they involve road crossings. That's not just important when you're talking to a second grader. If you want people to care about conservation, they need to understand it—which often means really being able to identify with the issue, to picture it and also relate to it emotionally. To *connect* to it. We've all seen roadkill. And we've all had to navigate from point A to point B. "Corridors make sense to people," said Nick Haddad, an ecologist at Michigan State University and the W.K. Kellogg Biological Station who is a leader in connectivity science. "So the greatest way we can accumulate conservation areas most rapidly is through a focus on corridors."

Thinking about corridors involves trying to see the world from the point of view of another creature. How can we make the human-dominated planet more welcoming for the other species that need to move across it? It was Andrew Gonzalez, the McGill ecologist, who really got me started trying to see the world through non-human glasses. As an undergrad at the University of Nottingham in England, Gonzalez took an ecology course and became fascinated by the idea that you could

see landscape fragmentation from space. He was intrigued by
the theoretical and mathematical approaches you could use
to study habitat fragmentation, and also by the pressing, real-
world challenge of solving the problem. It was the early 1990s,
and not much research had been devoted yet to understanding,
as he put it, "how entire communities and ecosystems might
undergo cascades of extinctions," and what would happen to
those ecosystems as a result. He began with an honors thesis
project, studying not forest fragments but far smaller ecosystems: patches of moss.

"Lumps of moss form little islands," Gonzalez, who now
holds the Liber Ero Chair in Conservation Biology at McGill
University, explained to me several years ago, when I visited
him in Montreal for a story I was writing. "The moss patches
are like a miniature rainforest; they're fascinating." The patches
teemed with tiny life, hundreds of different species of little
invertebrates. For several months, Gonzalez tested island biogeography theory, studying whether bigger islands of mosses
contained more species, and if you cut islands in half, whether
the loss of connectivity triggered extinctions. It was the SLOSS
question, writ tiny.

As a PhD student, Gonzalez returned to the moss, testing
whether linking the patches with corridors could slow extinction. He built corridors between some patches, left some isolated, and broke the corridors between others. He found that
the corridors did indeed stave off species' demise. In a later
experiment, published in 2011, Gonzalez and his McGill colleague Graham Bell set out to learn whether baker's yeast
could evolve to live in saltier conditions. They homed in on two
genes that played a large role in the yeast's ability to tolerate
salt, then tracked more than 2,000 populations of yeast over
scores of generations. They found that over a relatively short
period of time, the yeast could evolve responses to dealing

with a major environmental change. But a population's success in a saltier world depended on both the speed of the environmental change and on whether it was connected to other populations, enabling genes to migrate. Populations that experienced a slow increase in salinity and built up useful genetic mutations—a phenomenon known as "evolutionary rescue"—were far better able to survive a sudden increase later. Those mutations could spread into other populations, and save them, if the populations were connected.

When I met with Gonzalez in Montreal, he was busy assembling maps of which patches of land a bear, a nuthatch, a marten, or a frog can and cannot use—or "how each creature perceives the landscape," as he put it. Gonzalez and his colleagues built a computer model that took into account the way a network of forested areas surrounding the city of Montreal, the St. Lawrence Lowlands, looked to fourteen different species chosen to represent the region's vertebrates. Gonzalez was concerned about how urbanization was fragmenting habitat, and he hoped his research could help protect and restore linked areas of green space around and even within cities like his own.

— – – —

Back in the 1600s, what is now the urban metropolis of Montreal was a vast forest teeming with wildlife—including deer, moose, and beaver—whose presence was partly what enabled humans to thrive there. In fact, according to environmental historian Peter Alagona, that historical status as a thriving biodiversity hotspot is a common theme among cities. "Many of the biggest cities in the United States are located on sites that, prior to their founding, were unusually biologically diverse and productive compared with their surrounding regions," Alagona wrote in his book *The Accidental Ecosystem: People and Wildlife in American Cities.* "They were also crawling with wildlife."

It's true of Manhattan, too. When the English sea captain Henry Hudson first landed on the island in 1609, what he found was a place "with more ecological communities per acre than Yellowstone, more native plant species per acre than Yosemite, and more birds than the Great Smoky Mountains National Park," wrote ecologist Eric W. Sanderson in *Mannahatta: A Natural History of New York City*. If that landscape existed today, Sanderson wrote, "it would be the crowning glory of American national parks."

But Mannahatta does not exist today in any shred or semblance of its former self; nor does Montreal. Today, most non-human species cannot cross—much less inhabit—these concrete islands. Even far outside Montreal's urban core, forests and wetlands are giving way to suburban and exurban sprawl. The same is true in fast-growing cities around the world. Urbanization is one of the most powerful forces at work today. In 1950, 751 million people, or nearly a third of the global population, lived in cities. By 2018, that number was 4.2 billion people, or roughly half the world's population. By 2050, according to UN projections, more than two-thirds of the world's population will live in urban areas.

One of Gonzalez's maps showed areas of the St. Lawrence Lowlands that contained the right kind of habitat for ovenbirds—migratory warblers that breed in eastern North America and winter as far south as Venezuela—and how connected sections of that habitat were to one another. Ovenbirds eat insects, which they typically find by rustling through piles of leaves on the forest floor. That type of leaf litter builds up over time, and there is more of it in older forests with larger trees. "There's a

An ovenbird searches for nest-building treasure. Ecologist Andrew Gonzalez studied how urbanization was fragmenting habitat for ovenbirds and other species outside Montreal.
Marie Read/NPL/Minden Pictures

very clear corridor to the northeast of the city—large woodlots, pristine forests, biosphere reserves," Gonzalez told me, as we pored over a map at a coffee shop on a rainy Montreal morning. The map was colored in varying shades of gray, with a series of red and yellow splotches, some of which formed a band. Those colored patches represented the most important habitat for the ovenbird. The corridor Gonzalez was talking about was made of areas that glowed red on the map. Much of it wasn't necessarily contiguous forest, but it was a region of stepping-stone patches farthest from the city's suburban development. Conserving this land was crucial to the bird's survival.

The research showed that protecting just the most important 12 percent of pixels on the map would, in fact, protect 57 percent of the bird's existing habitat. That's a pretty good return on a relatively small conservation investment. Overall, for the species Gonzalez used in the model, protecting the most essential 17 percent of the remaining forest would preserve nearly three-quarters of the region's connectivity. With foresight and smart planning, a little bit of effort can go a long way. Without any action to protect or reconnect these areas, though, by 2050 the forest cover will have shrunk by another 12 percent, with fewer, smaller, less-connected patches of habitat. By that point, it will have become much more difficult to undo development and protect the species occupying the scant forests that remain.

One reason corridors can be so effective for conservation is that they aggregate. Individual corridor projects might be relatively small-scale or hyper-local, such as turning your backyard into a pollinator garden, or replacing your lawn with native plants that provide food and shelter to a host of birds and bugs. Now imagine if everyone on your block did the same—or everyone in your neighborhood. In the contiguous US, according to research, "single-unit detached housing and its associated private land use" occupies nearly a third of the country's land area.

That's a lot of potential. In suburban and urban areas, back-yards might be relatively small, but there are zillions of them. As habitats, they might appear isolated or fragmented, but for birds especially, a corridor made of yard-size native plant patches, filled with tasty seeds and insects, a water source, and a place to hide or nest, just might be the difference between life and death.

Collectively, corridors can add up to something huge. Mixing in wildflowers, hedgerows, or strips of prairie along agricul-tural fields can create corridors for insects that pollinate crops and other plants, and habitat for birds. They might not seem especially exciting, but if every farmer in a region planted these, the impact could be transformational: Farmland cov-ers roughly two-fifths of the US. Even simply planting a mix of different crops in a mix of different field sizes can make a big difference in terms of what species can move through the landscape, said Claire Kremen, an ecologist at the University of British Columbia who runs a lab called Working to Restore Connectivity and Sustainability. Kremen has spent much of her career studying how to make agriculture more biodiversity-friendly, including how creating corridors of native plants on, across, or adjacent to farm fields helps attract native bees that can pollinate crops. You can't have "intensive nature and inten-sive agriculture," Kremen said. "Separating people from nature is outdated."

Haddad, the Michigan connectivity scientist, is also studying corridors on farms, and believes they can be a powerful addi-tion to a network of corridors. If you conserve prairie strips on farms, along with vegetation along the banks of streams and rivers, add in urban greenways, maybe even some moun-tain ranges, "soon you have corridors going across the planet," Haddad said. In 1994, as a graduate student at the University of Georgia, Haddad set up a corridor experiment. The Savannah

River Site Corridor Project, now in its fourth decade, sits in a seemingly unlikely place: a restricted-access former nuclear-bomb-building site owned by the federal government.

In the early 1950s, the US Atomic Energy Commission (the precursor to the Department of Energy) built the Savannah River Site (SRS) near Aiken, South Carolina, to produce tritium and plutonium-239 for use in nuclear weapons. To house the bomb factory and all the related toxic and radioactive waste storage, and surround it with necessary buffer zones, the government needed more than 300 square miles. So it relocated an entire town of 6,000 people, graveyards and all. "It's hard to understand why our town must be destroyed to make a bomb that will destroy someone else's town that they love as much as we love ours," read a handwritten sign that one resident tacked up in Ellenton, South Carolina, shortly after the government announced its plans. An early photo from the corridor experiment shows a billboard at one entrance to the government site that reads, "Starve a spy, feed a shredder." It doesn't exactly look like a place that would draw ecologists.

But when the Atomic Energy Commission first launched the weapons facility, it also requested an inventory of the site's biological diversity and environmental conditions, to use as a baseline before the radioactive work began. The government hired University of Georgia ecologist Eugene Odum, considered "the father of modern ecology," to conduct the survey. That initial research launched the Savannah River Ecology Laboratory, a division of the university that today employs researchers studying all kinds of ecological and biological processes at the site.

When Haddad started his PhD at the University of Georgia, he was interested in habitat fragmentation. He also knew that the forest service was planting and logging trees. Before Europeans settled the area where SRS sits today, Indigenous

people—including the Westo, Edisto Natchez-Kusso, and Yamassee people—had managed the longleaf pine savanna using fire. By 1950, though, most of the savanna had been cleared for agriculture. Across the Southeastern Coastal Plain, stretching from North Carolina to Louisiana, nearly three-quarters of the natural ecosystems, and as much as 97 percent of the longleaf pine landscapes, are gone. When the weapons site was built, the US Forest Service launched a colossal reforestation effort, planting 40,000 trees a day for the first two years and 100 million of them by 1968. Initially, the Forest Service planted loblolly and slash pine to build a border wall between the edge of the site and the nuclear reactor. Today the site is largely forested again, with a million trees a year harvested for timber and then replaced.

Haddad thought perhaps the Forest Service might be willing to fell trees in a way that could align with research. They quickly agreed to collaborate—but they didn't like Haddad's study design. He was interested in how habitat fragmentation impacted populations of butterflies, and the map he'd created of a potential project "looked like if I took a shotgun and sprayed shots across the SRS," he recalled. "It made total sense for me thinking about spatial patterns of fragments," but the Forest Service saw no practical application. "They came back with the idea of corridors." Haddad was enthusiastic.

When he and fellow grad student Robert Cheney went in to meet with the agency to talk about a budget, Haddad was thinking small. He hoped they might give him $10,000, which seemed like more than enough for a PhD project. But Cheney had another number in mind. He'd worked in industry and on big restoration projects for The Nature Conservancy. Before Haddad could speak, Cheney asked for $50,000. The Forest Service manager said yes. And that was just for the project's first year. The agency also funded four other labs—and within a couple of years, things were humming along. The projects in

that first decade were all about dispersal—the process by which animals and plants spread out through a landscape. Haddad was studying butterflies, Douglas Levey's lab at the University of Florida was working on how birds spread the seeds of fruiting plants, and other labs were working on bee pollination and the dispersal of small mammals.

Ecologists use the term "matrix" to refer to the mix of landscape types that animals must navigate outside of protected habitat—it might be ranches, farms, villages, backyards, freeways, strip malls, and all of the above. At other corridor experiments, like Lovejoy's, the corridors are composed of forest, and the matrix is land where the forest has been cleared. SRS is designed so that open patches of regenerating longleaf savanna are surrounded by the matrix of dense pine forest. While the corridor-matrix makeup is flipped, the concept is the same. "The corridors and patches are the habitat that is appropriate to the study region, in South Carolina and Brazil," said Haddad. "Suitable habitat," longleaf pine savanna or tropical forest, is surrounded by "unsuitable habitat," dense pine plantation or agriculture.

At SRS, some of the open patches are connected via corridors, so from the air they resemble cartoon barbells, two squares joined by a 150-meter-long pathway (about 164 yards). Others are isolated, either in individual squares surrounded by forest or in squares with "wings"—corridor-like pathways that stick out from the squares but don't connect them to any others.

Once the experimental floodgates were open, researchers came rushing in to study all kinds of things related to corridors.

Monarch butterflies can migrate up to 3,000 miles, from southern Canada or the northeastern US to Mexico. Across the continental US, in the first two decades of the twenty-first century, butterfly numbers declined by more than 20 percent.
Brendan George Ko

The project began with a question: "Are these things real super-highways for plants and animals?" The answer was a definitive "yes," said Haddad. "And in surprising ways." The overarching results of the Savannah River Site Corridor Project show that corridor effects increase over time; they still haven't leveled off at the experiment site. Connected patches of habitat there house 20 percent more plant species, and they have higher rates of new plants coming in and taking root, and lower rates of plants going extinct.

I visited SRS early in the summer of 2021, when the world was starting to reopen on the heels of Covid vaccinations. About a dozen people convened in Aiken, including Haddad and other scientists who'd been working at the site for years, as well as some grad students just starting their research, for the team's annual meeting—which begins, by tradition, with a picnic followed by dessert at a popular downtown Aiken ice cream shop called Flanigan's. Before six o'clock the next morning, we were on our way to the Department of Energy badge office; accessing the corridor facility involves accessing the former nuclear weapons site, so it's nothing like other just-walk-into-the-woods ecology study areas.

Ellen Damschen, an ecologist at the University of Wisconsin–Madison who has been running experiments at SRS for two decades, led the group through the field sites. Damschen spends long days meticulously chronicling the plants growing in the corridors, traipsing through fields of poison oak. Like much of the work at SRS, Damschen's research looks at the impact of corridors on plants rather than animals. Since plants equal habitat and food for animals, the two are intricately linked. She and her frequent collaborator Lars Brudvig, an ecologist at Michigan State University whose work focuses on restoring degraded landscapes, can name an astonishing number of plants that grow in their study area—legumes, grasses, sedges,

and so on. For much of their careers, they've been restoring the longleaf pine savanna, watching plants slowly colonize new areas via the corridors.

During fieldwork weeks, Damschen and Brudvig will be out at their sites for twelve hours at a time, walking back and forth, row by row, across squares of land each a little smaller than a tennis court. "We both like endurance sports," Damschen joked.

A decade ago, Damschen designed an experiment where scientists mimicked the way wind disperses seeds, so they could track whether corridors made a difference. Scientists created artificial seeds out of yarn dusted with fluorescent powder, released them into the wind via boxes mounted on poles, and searched for them in the dark using black lights. The study showed that corridors can influence the way wind disperses seeds, and that when the corridors were aligned with the direction of the wind, more seeds could move longer distances. The research added to the growing pile of corridors-as-superhighways evidence: Not only were corridors helping plants and animals disperse on the ground, but they were even helping plants disperse on the wind.

More recently, Damschen led a team of SRS scientists in synthesizing nearly twenty years of data from the site. They found that for 239 species of plants, the rates of colonization each year were 5 percent higher in corridors than in fragments that weren't connected. The corridor experiment has also shown that connectivity promotes "spillover," spreading biodiversity beyond the boundaries of a protected area. In research published in 2009, Damschen, Brudvig, and other SRS researchers found that there were more plant species around patches of habitat that were connected by corridors than in ones that were isolated.

Trying to link up existing protected areas via corridors is a major international goal. A study published in 2022 set out to

map connectivity among the world's protected areas, in terms of the ability of mammals to move among them. (The study didn't distinguish between the type or size of the animals. It was based on a model that uses the flow of electrical currents to approximate mammal movement.) That study found that making human landscapes between protected areas more welcoming to wildlife—what the researchers called "mitigating the human footprint"—would improve connectivity more than creating new protected areas (although doing both was the best plan of all).

— – – —

Part of what lies between these protected areas, as my seven-year-old understood that night while reading about aliens, is *roads*. Roads are the biggest contributors to fragmentation and are seriously deadly to wildlife. By extremely conservative estimates, 100 million large mammals are killed around the world each year by vehicles on roads. That's just large mammals. It doesn't account for small mammals, not to mention birds, snakes, turtles, frogs, crabs, butterflies, and hundreds if not thousands of other types of critters across the globe, some of which are already endangered. "The amount of endangered species being hit by the side of the road is staggering and unaccounted for," said Gary Tabor, founder and CEO of the Center for Large Landscape Conservation, a nonprofit based in Montana.

I first met Tabor on a plane to Alaska in the early 2000s. At that point, he'd been working on connectivity issues for nearly two decades already—and he's been at it ever since. He views it as a health issue. "Connectivity is the circulatory system of nature, and we've done a really poor job of conserving it," he said.

Researchers at the Savannah River Site Corridor Project set up traps to collect seeds dispersed by wind or birds. Years of studies at SRS have shown that corridors increase seed dispersal. *Nick Haddad*

"We've focused on the heart and lungs but not the system." The heart and lungs, in this case, are the areas around the world that people have set aside as off-limits to development. But, as Tabor put it, "those places can't survive unless we've protected the circulatory system." Nor can those places "be a resilient system to buffer climate change."

Of the roadkill crisis, Tabor told me, "I don't think people understand the magnitude of this. They really just think of it as, 'I saw a dead animal by the side of the road.'" That 100 million estimate also only accounts for "direct mortality." It doesn't include the untold gazillions of animals killed because of the ecological changes wrought by roads. But roads "can open a Pandora's box of environmental problems," warned a team of scientists in 2014. Led by the Australian ecologist William F. Laurance, they proposed a "global roadmap" that could guide road building around the world to ensure the greatest human benefits with the least cost to the rest of the planet's life. Especially when roads plow through areas that are essentially wilderness, they "often dramatically increase land colonization, habitat disruption, and overexploitation of wildlife and natural resources," the scientists wrote in the high-profile journal *Nature*. What's more, they said, the deforestation we create by constructing roads is actually "contagious," since "new roads tend to spawn networks of secondary and tertiary roads that greatly increase the extent of environmental damage"—habitat loss, poaching, illegal mining, destabilizing invasions of nonnative species, wildfires.

At the time, the International Energy Agency estimated that between 2010 and 2050, "global passenger and freight travel" would double, with the vast majority of that growth happening in the developing world. A 2013 study by the agency found that to meet that surge in travel demand, people would need to build roughly 25 million kilometers (more than 15 million miles) of

paved roads and 335,000 kilometers (about 208,000 miles) of railroad tracks by the middle of the century—an increase of 60 percent. The additional roads alone would be "enough to encircle the planet more than 600 times," according to another op-ed Laurance published in the journal *Nature*. During the first ten years of the twenty-first century, China had already increased its paved roads and railroad tracks by 290 percent, according to the same report. In 2000, 53 percent of the world's roads were paved. By 2010, it had risen to 60 percent.

The total number of roads that might crisscross the planet in 2050 would be 1,000 times the circumference of the Earth. Put another way, if you drove at 100 kilometers an hour, or about 62 miles per hour, it would take you forty-five years of nonstop driving to cover all of them. Yet paved roads as we know them today, made of asphalt and concrete, have only existed for about a century. This is a radical, almost incomprehensible amount of change in the blink of an eye.

All those additional vehicles driving on all those additional roads would also need a place to park. The report estimated that from 2010 to 2050, parking space would increase from 30,000 square kilometers (nearly 12,000 square miles) to 80,000 square kilometers (around 30,000 square miles). "This addition equates to nearly the size of Costa Rica in total area," the report said. And that's a conservative estimate. If developing countries built parking spaces as big as those in the US, yet another 30,000 square kilometers of the planet would be paved over. For parking.

A decade later, the website Energy Monitor reported that $3.5 trillion had been invested in road construction projects that were underway in 2023. Across every part of the world, spending on roads dwarfed spending on any other forms of transportation, like public transit systems. Many of these roads are being constructed in ecologically sensitive landscapes—including

across areas officially recognized as crucial ecosystems in need of protection, and even across existing national parks.

One of the biggest drivers of ecologically catastrophic new paved roads is a massive Chinese effort called the Belt and Road Initiative—a multi-trillion-dollar project involving highways, railways, dams, pipelines, and other infrastructure across 64 countries, largely in Africa and Asia but extending even into Western Europe and South and Central America. It's billed as "the largest infrastructure project in human history." And it could spell disaster for wildlife populations. A 2017 report by the conservation group WWF found that proposed Belt and Road projects overlapped with the ranges of 39 critically endangered species and another 81 endangered ones, as well as another 145 threatened species. The projects also interfered with more than 1,700 areas internationally recognized as vital to biodiversity. And the report predicted that, in the Belt and Road countries, more than 30 percent of protected areas could be impacted.

In 2019, one Belt and Road project slapped a railroad line straight through Nairobi National Park, bisecting habitat for iconic wildlife like zebras, giraffes, and leopards. Thanks to fierce opposition, the park portion of the railroad line is elevated rather than on the ground. But it nonetheless impacts wildlife whose habitat is already shrinking, and it could just as easily have gone around the park instead of through it.

In perhaps the most devastating project so far, the Pan Borneo Highway, currently under construction, is cutting across habitat that is crucial for hundreds of species including orangutans, sun bears, pygmy elephants, and clouded leopards, in an area also home to dozens of Indigenous communities whose livelihood will be destroyed. Beyond simply the linear barrier of the road, the highway—which cuts through an area called the Heart of Borneo that contains some of the island's

last remaining old-growth forest—is opening up adjacent areas to industrial development.

Road building can be done smarter. It could follow the ecological roadmap that Laurance proposed. It just takes will. It also takes, in many cases, bridges and tunnels created for animals. Wildlife crossing structures, like my son was referring to—overpasses and underpasses that help animals safely get from one side of a road to another—offer a clear, proven solution to the problem of roads as barriers to wildlife. And they're not just good for animals; they can reduce vehicle collisions with animals by as much as 97 percent, enabling animals to cross even a major interstate highway and allowing connectivity between the landscapes on either side. As one report put it, "Unlike many large-scale problems facing society today, there are proven solutions to reduce wildlife-vehicle collisions and reweave native habitats."

Luckily, wildlife crossings are proliferating across the planet, and particularly around the US, as communities and transportation planners increasingly realize their benefits for both wildlife and humans. Colliding with a deer, after all, is bad for everybody involved. In the western US, from 2021 to 2023, eighteen wildlife corridor and crossing laws were passed, with multiple states appropriating tens and even hundreds of millions of dollars for overpasses, underpasses, and other wildlife-friendly infrastructure. The federal Bipartisan Infrastructure Law, passed in 2021, included $350 million over five years specifically for projects that reduce wildlife-vehicle collisions and increase connectivity, with 60 percent of the funding intended

Next Spread: Wildlife overpass in Dwingelderveld National Park, the Netherlands. These types of structures are increasingly popular—and highly effective—tools for helping animals safely cross dangerous roadways. *Rudmer Zwerver/ Shutterstock*

for rural areas. The Federal Highway Administration received sixty-seven applications from thirty-four states for its Wildlife Crossings Pilot Program, with requests totaling $549 million— five times the amount of funding available for the first year.

Around the world, wildlife crossing structures have proved incredibly successful, and many more are planned. In just one example in Kenya, a narrow, fenced corridor built in 2010 specifically for elephants connects Mount Kenya National Park and the indigenous Ngare Ndare Forest Reserve, enabling passage for elephants along a much larger migration route. Eight and a half miles long but only a few hundred feet wide at its narrowest point, the Mount Kenya corridor provides safe passage for elephants across farmland where they would otherwise raid crops, causing conflict with humans. At one point the route crosses a highway, so the corridor includes an elephant underpass. "We were very cynical it would work," Lucy King, head of the human-elephant coexistence program for Save the Elephants, recalled. "But within the first twenty-four hours, the elephants went through it." Driving that road once, King decided to stop her car and get out—and there below her in the grass was an elephant using the underpass.

Wildlife crossings are mostly uncontroversial and apolitical. What's not to like about stopping cars from colliding with animals? Even if you care only about people, they are, for the most part, a relatively easy sell—albeit a sometimes expensive one. In the US, an underpass can cost from about $500,000 to nearly $3 million, and an overpass can run more than $6 million—a drop in the bucket, though, when you consider that across the western US alone, collisions with large mammals cost more than $1.6 billion a year.

Of course, there are some outliers; some places just like to make their projects splashier. As I write this, the most celebrated wildlife crossing under construction costs an astronomical $90

million, and unsurprisingly, it's in Los Angeles. The Wallis Annenberg Wildlife Crossing spans the eight-lane Ventura Freeway, plus two breakdown lanes, and an adjacent frontage road to provide safe crossing and restored connectivity for mountain lions, bobcats, deer, lizards—even birds.

Most people can easily grasp the idea of linear connectivity on land, but at first glance it's hard to understand with birds. Can't they just fly to the next place? In fact, lots of discussions about connectivity explicitly focus on "non-flying" animals. But linear infrastructure can actually work as a physical barrier for flying animals as well. The wrentit, a little songbird that's native to California, won't fly across that stretch of the Ventura Freeway. Research published in 2017 found distinct genetic differences among wrentit populations on either side of the freeway, showing that the road presented a physical barrier for the birds. The Wallis Annenberg crossing will also help the wrentits, whose preferred habitat of coastal sage scrub and chaparral will make up the overpass's vegetation.

It's true that many birds can simply fly over roads, fences, or border walls, but that is only one aspect of connectivity. There are other challenges. Birds' life cycles often involve multiple habitats, sometimes hundreds or thousands of miles apart, and if something is amiss in one of those places, things can go very wrong in another; places are connected in ways that humans can't necessarily see. For migratory birds, this is especially true—and some birds' migration journeys involve habitats in locales as geographically and politically distinct from one another as Alaska and Chile (Hudsonian godwits), or Siberia and Texas (sandhill cranes). In one story that's become a cautionary tale, ornithologists in the mid-1990s had noticed that the Swainson's hawk, a large raptor that lives in grasslands, was declining across the western US and Canada, where it hunts rodents and times its breeding to coincide with grasshopper

Swainson's hawks migrate across two continents, flying
from Alaska to Argentina and back in a year, a round-trip
equal to two-thirds of the Earth's circumference. These
large grassland raptors need food and habitat across that
entire journey. In the 1990s, researchers discovered that the
birds, which eat insects when not breeding, were dying from
pesticides that Argentine farmers were spraying to prevent
grasshoppers from ruining their crops. *Christina Speed*

outbreaks to feed its young. (The species was listed as threatened under California state law in 1983, but not listed under
the federal Endangered Species Act.) Birds would leave in the
fall, but many of them would not return in the spring. No one
knew much about the birds' migration routes until one biologist
used one of the earliest satellite tracking devices small enough
for birds and tracked a Swainson's hawk from California to
Argentina. There, in the pampas a few hundred miles west of
Buenos Aires, he "found huge flocks of hawks soaring over
alfalfa and sunflower fields during the day and roosting at night
in eucalyptus groves planted near ranch buildings," according
to a 1996 story in *The New York Times*.

The biologist, Brian Woodbridge, "also found hundreds of
dead hawks, and a farmer told him the birds had died after a field
next to their roost was sprayed," the *Times* story recounted. It
turned out the Argentine farmers were controlling grasshopper
outbreaks with an extremely toxic pesticide. Woodbridge and a
grad student counted a shocking 4,000 dead Swainson's hawks
"and estimated that 20,000 of the raptors had died from aerial
and ground spraying of a pesticide that forensic evidence later
confirmed to be monocrotophos," according to the story. "Some
of the birds were killed by direct exposure to the spray, and others by eating poisoned grasshoppers, Mr. Woodbridge said."

The American Bird Conservancy, a nonprofit, ultimately
convinced Ciba-Geigy, a Swiss company, to remove the pesticide from the market in Argentina (it had never been approved
for use in the US) and work with other manufacturers to do the
same. Since then, Swainson's hawk numbers have stabilized,
according to the American Bird Conservancy, but loss of habitat remains an issue.

The Swainson's hawk story, which I first heard from a nat-
uralist on the Gulf Coast of Texas, illustrates the astounding
nature of the connectivity issue for birds—and how difficult

conservation can be, given that it has to encompass every part of a bird's habitat throughout the year. A "corridor" for a bird can stretch across continents, connecting a sunflower farm in western Argentina to a prairie in western Canada. Swainson's hawks migrate roughly 12,000 miles round-trip each year. "What does conservation look like when you have birds that cross states, international borders, continents?" mused Stan Senner, who at the time was vice president for bird conservation at the National Audubon Society, when I called to chat one winter afternoon. "How does one protect birds like that?" And a problem for birds in one migration stop can ripple across the corridor, affecting ecological communities everywhere else they go.

Compounding the problems, birds' exact flyways, and the places where they stop to rest and feed along the way, are in many cases poorly understood. "Bird banding records were the original way we learned about migratory routes," said Autumn-Lynn Harrison, a research ecologist at the Smithsonian Conservation Biology Institute's Migratory Bird Center, in Washington, DC. But that science relied on chance—someone in another location had to see and catch a bird that has been banded elsewhere. Recapture rates of small songbirds in a different place from where they were banded, for instance, are less than 1 in 10,000.

One of the birds Harrison studies is the black-bellied plover, a ground-nesting bird that breeds in the Arctic tundra, laying eggs she describes as "the color of lichen." Over 100 years of

Eggs of the black-bellied plover. These long-distance migrants rely on wetlands, prairie potholes, estuaries, and tidal flats and can vary their routes by more than 300 miles (500 km) in a year, due to changing water levels. Ensuring connectivity for birds like this requires regional-scale conservation.
Ken Russell/Laurel Devaney/Alaska Region U.S. Fish & Wildlife Service

observing banded black-bellied plovers, just two individuals were ever seen a second time. Yet those two birds helped scientists connect breeding to wintering habitat, and prompted Harrison to start a tracking project to study birds from multiple breeding colonies. "We didn't know the connections between sites, which populations go where, how do they get from point A to point B," she said.

Harrison outfits the birds with tiny harnesses carrying tracking devices that weigh less than five grams. Her research has tracked some birds for up to seven years and found that they have "high fidelity to the same breeding location and wintering site—within a kilometer of their location the previous year." But the migration routes they take, stopping in the prairie potholes of Saskatchewan, Alberta, and North Dakota for up to two weeks, can vary by as much as 500 kilometers (300 miles). "They have to stop where the resources are," said Harrison. Water levels can vary dramatically from year to year, and the birds might not have available water in the places it existed the year or decade before. This means that birds like the black-bellied plovers need multiple options to stop along the way, and we need regional-scale conservation measures to ensure "a regional matrix of potential areas," Harrison said.

In the spring of 2022, Harrison, like many of her fellow ornithologists, was eagerly awaiting a game-changing migration-tracking technology: a shipment of new, lightweight, tiny tags developed at the Max Planck Institute in Germany. The tags, which had been in development for a decade, are part of a new tracking system called Icarus. Harrison's colleague Amy Scarpignato, another researcher at the Migratory Bird Center, was also awaiting Icarus tags for summer fieldwork that year. Scarpignato had hoped to receive ten new Icarus tags to study eastern meadowlarks, a grassland bird whose numbers have been declining, thanks to the steep decline in native prairie

habitat, and whose migration routes also remain a mystery. The tags, which were predicted to greatly expand the kinds of birds researchers could track, were designed to connect not with a satellite but with a module on the International Space Station (ISS). There was just one problem: It was on the Russian side.

The war in Ukraine and its resulting political fallout had left Russia isolated from much of the international community, and the country had brought down an iron curtain on its side of the ISS, rendering its equipment off-limits to American and European scientists. Bird conservation had gotten tangled up in geopolitics, leaving another corridor mystery to solve later.

— – – —

Linear connectivity is essential to thriving wildlife populations. It's simple and straightforward, though that doesn't necessarily mean it's easy. But without it, we face a lonely world, overrun by our own kind and the animals we've bred for food. At the same time, it's becoming increasingly clear that the matrix is crucial to protecting biodiversity. "Conservation is catching up to this idea that protected area systems aren't going to do the job," said Zia Mehrabi, an ecologist and data scientist at the University of Colorado Boulder. "The reason it doesn't work is that animals move. All kinds of things move through landscapes." As human population continues to climb upwards of 8.1 billion, and as climate change shifts the areas where the planet's millions of other species can live, wildlife cannot survive without using places that we may like to think of as human habitat. This makes corridors an increasingly vital conservation tool.

Still, it's also true that a focus on linear corridors is an imperfect attempt to superimpose a specific mindset on the natural world—one that's been molded by both colonial attitudes and a contemporary car-centric point of view. "We have designed our world around transportation engineering," said David Theobald, a landscape ecologist and conservation scientist

based in Colorado. Scientists and conservation professionals, said Theobald, like to think about linear corridors because they represent "defined units" within a landscape, meaning they are easy for humans to envision—and to manipulate. It's based partly in a need—among humans generally and urban planners in particular—to say, as Theobald put it, "Here's the line, and we'll do something here and not there."

On an outing with Theobald one summer morning, we stopped at an intersection just north of Boulder, where US 36, a two-lane highway, winds its way along the base of the foothills toward Colorado's Rocky Mountain National Park. We watched a hawk scouting the ground from a power line and talked about the wildlife we'd each seen moving through this area over the years: elk, deer, raccoons, coyotes, skunks, prairie dogs, rabbits. Animals moving between the foothills and the plains must cross this highway. I told Theobald that I used to volunteer at a wildlife sanctuary up the road, recalling how nervous I often was as I drove home from my shifts on summer evenings, always concerned I might hit some poor creature that, like me, was simply trying to get somewhere safe for the night. "I just saw some roadkill back there that looked like a weasel," Theobald said.

"Before we were here," he said ("we" meaning modern asphalt-obsessed people), "this was just a big, open, permeable area." A wildlife corridor here, in the form of an overpass or underpass, might be a step forward—but it could also be seen as an overly simplistic way to fix linear problems with more lines. "Corridors are a construct for us to think about, to draw it on the board really easily," Theobald said. We need to also think about the bigger picture—what connectivity scientists call "landscape permeability." This term is less about a specific corridor through which one species can move between places, and more about the general structure of an area. What barriers does it contain that might make it more or less usable for

wildlife? Thinking about permeability helps shift the emphasis to restoring whole landscapes instead of simply building corridors across them. It involves thinking about how the puzzle pieces fit together in a broader sense. "How does north Boulder connect to Rocky Mountain National Park and to the southern Rockies?" Theobald asked. "There is value in connecting mountains to plains."

Still, it was hard to wrap my head around possible solutions, besides building ways for animals to go over or under the highway. "How else could you really fix this?" I asked, as we stood on the east side of the highway, facing west toward the mountains. "You can't take the road away." Theobald didn't immediately reply, so I tried again. "Right? So what can you do, besides create overpasses, underpasses? Maybe thinking about it in that linear way is a fix."

"It's a pragmatic approach," he said finally.

"But what else could you do, other than take away the road?"

He smiled. "That's one option!" He paused again. "Why do we have to go around so busily all the time? If Covid didn't expose that, nothing would. So that's one option, and I think we need to have discussions about that. I think we need to think outside of the last 100-year trends."

It was an enticing idea. What else might our world look like? So much of our lives, our assumptions, our mindsets have been shaped by the need and desire for roads and more roads—wider, longer, cutting across more and more places. What if we envision a different future?

Around the world, projects are underway to make our human habitat more welcoming to wildlife, and to reduce the kinds of conflicts that can result from sharing limited and progressively more-chopped-up space. The matrix seemed to be an important concept, and I wanted to learn more. Luckily, I knew just where to go. I booked a flight to Central America.

CHAPTER 3   *Costa Rica's Osa Peninsula*

The Matrix: Creating Climate Corridors So Animals Can Move

As our Cessna Grand Caravan turboprop dipped over the shimmering waters of Costa Rica's Golfo Dulce in the rainy season of 2022, I couldn't wait to get back into the forests of the Osa Peninsula. More than twenty years earlier, this spit of land in the country's southwest was the place where I had first experienced a tropical rainforest. I could still remember the jitters I'd felt, a mix of excitement, nerves, and motion sickness, as the little plane made a U-turn in the sky before touching down on the airstrip beside the cemetery in Puerto Jiménez. Back in 2000, Costa Rica was becoming an increasingly popular tourist destination, but most of its eco-lodges were still fairly rustic affairs, particularly on the Osa.

Previous Spread: Farmland in southern Costa Rica, near the Talamanca Mountains—a vast protected wilderness area that contains one of the largest forests remaining in Central America. Connecting those mountains with two national parks, across miles of farmland and human infrastructure, is crucial for protecting wildlife pathways. *Rodrigo Benavides*

Shortly before that trip, I'd left an editing job to become a freelance writer. I had no cash to spare, but I'd secured an assignment to write about an upscale eco-resort for a new online travel publication. It wasn't exactly a daring adventure, but it was my first real nature-based assignment, and I was high on the dizzying new possibilities that online magazines promised for journalists (promises that didn't really pan out). Mostly, though, I was thrilled that my dream of visiting a rainforest was about to come true. By the time I went to bed that first day, I had seen three species of monkeys, scarlet macaws and toucans and trogons, iguanas, coatis, blue morpho butterflies, and a host of other tropical creatures that filled me with awe and a strange emotional mix of joy and despair.

Most people who have visited a tropical forest will recognize this churning ball of feelings. One part is gratitude for a world in which such beauty exists in so many countless forms—a mated-for-life pair of macaws swooping across a valley; a line of leafcutter ants lugging their carefully sliced morsels of vegetation across the forest floor to their underground gardens; an algae-coated sloth hanging by her toes, a baby tucked to her belly and a village of bugs and fungi frolicking in her fur; a tree covered in spikes that may have once protected it from the sloth's giant predecessors, or might be an adaptation for fending off strangler fig vines. The other part of the feeling is an excruciating understanding of loss—of creatures and their habitats, of the boundless natural systems where they once thrived, of the world they inhabited before capitalism and rampant extraction upended everything.

Today, the eco-resort where I stayed costs around $1,200 per person per night, and it is just one among numerous luxury lodges, retreats, and expat holiday homes that dot the Osa. Yet compared to much of Costa Rica, this peninsula can still feel like the outback.

A couple of years after my first trip, I returned to the Osa with a group of ecologists; it was where I first witnessed the slow pace of fieldwork, and where I learned just how patient and persistent researchers need to be. From sunup until sundown, I trailed the scientists across a hillside as they studied the flow of minerals, nutrients, and gases through the forest. We collected leaf litter—the leaves and twigs and other bits of forest detritus lining the ground. We installed gas-exchange chambers in the soil, DIY contraptions we'd built out of sawed-off Tupperware. We did other tasks I can't now recall, involving test tubes and lots of kneeling in the dirt. We wore heavy snake chaps and tall rubber boots, and we wiped our sweat with increasingly soggy bandanas and swore at the biting ants.

Now I had come back because the Osa turns out to be a perfect place to think about wildlife, interconnected systems, and how we can ameliorate the matrix. It's home base for an international team of researchers who are thinking about the future of Central America's wildlife through the lens of "climate corridors," pathways that animals can use to access new habitat as the climate shifts. Before I could get back into the forest, though, I was headed to the highlands. I remained on the propellor plane as it landed by the little cemetery, deposited a few passengers, and took off again, headed for the other side of the Golfo Dulce, where a lanky, bearded ecologist from the north of England was waiting for me.

— – – —

Andy Whitworth has a dry sense of humor, a laid-back vibe that sometimes masks the urgency he feels about his work, and a

---

Andy Whitworth of Osa Conservation ascending a fig tree. Osa's staff and volunteers frequently climb huge rainforest trees to operate camera traps, build rope bridges, look for signs of wildlife, and otherwise monitor the forest. *Charlie Hamilton James*

soft spot for dogs. As executive director of Osa Conservation, a local nonprofit with an increasingly large mission, he does everything from climbing fifteen stories up in a tree to courting donors to capturing peccaries and fitting them with radio collars. Whitworth is more at home in a rainforest than anywhere else. At Osa's field station, his constant companion is Charapita, a black mutt he rescued and named for a hot pepper from the Peruvian Amazon. A second dog, a puppy Whitworth found drenched and shivering in a rainstorm, now lives at the station too, much to Charapita's dismay. As someone who lives with several rescue dogs, I felt right at home.

Whitworth grew up near Manchester, where his father, a former taxi driver, ran a pet store. "In this really industrial city life, with no wilderness, I was always around these exotic animals, like parrots," he recalled. The shop was mostly birds and fish, but one day his dad took him along to a friend's pet shop in a nearby town. While the two men were having tea, the seven-year-old Whitworth stuck his hand in a tank and grabbed a bunch of corn snakes. He was captivated (and thankfully not constricted). His dad let him begin keeping some snakes, and before long he was breeding snakes and frogs from around the world, a hobby he continued until college. But while he loved being around all the animals, he was aware from an early age that "there was something not quite right. I remember things coming in, new shipments. I didn't like the idea of things coming in from the wild in crates." That feeling prompted Whitworth to study zoology in college, and then to work for several years in a zoo, where he first learned about wildlife conservation.

In his early twenties, Whitworth told his father he did not want to take over the family business. His dad sold the shop. "It's still there," Whitworth said. "I think they mostly do birds and fish. I would imagine a lot of fish probably still come from

the wild. Which is why I have to suffer in the tropics for the rest of my life and do penance."

He was only partly joking about the suffering. For tropical biologists, parasites are par for the course. Whitworth has had four bouts of leishmaniasis—a potentially fatal parasitic infection that can result in skin sores and spleen damage—as well as botflies (fun little maggots that burrow under your skin) and other setbacks caused by creepy crawlies. But none of it has dimmed his love for the rainforest and just about all of its inhabitants. Over dinner one evening at the field station, when a large black beetle landed on our table, Whitworth eagerly picked it up. "You have to listen to it," he said, and began holding the beetle up to our dinner companions' ears. It emitted a ghostly noise, somewhere between singing and screaming. Setting the insect back on the table, Whitworth turned it over, belly up. Its underside crawled with mites and teensy arachnids called pseudoscorpions. Watching him swoon over the beetle and its stowaways, it wasn't hard to imagine the kid with his hand in the snake tank. I was glad we had already finished our meal.

Whitworth spent years working in the Peruvian Amazon, and when he first came to Costa Rica, he was struck by the difference in scale. "Manu National Park in southeast Peru is bigger than Wales, right," he explained in his characteristic manner of speaking, more like a friend at the pub than an esteemed biologist, as we wound our way uphill from the airport. "It's so big that it can probably self-function as an ecosystem, to some extent. But the parks here in Costa Rica aren't big enough for that. Almost all of them are too small to really hold healthy populations of the big stuff, of the things that we need to keep that ecosystem healthy." A study published in the journal *Science* just weeks earlier had reported that over the past 130,000 years, food webs for land mammals had collapsed. More than half the links in those webs have vanished. Food webs are an

essential part of how ecosystems function, and without all the components, those systems could be in trouble.

Many protected areas around the world are simply too small to meet the needs of species living there. Corcovado National Park, which occupies about a third of the peninsula and extends out into a protected area in the sea, is a crown jewel of Costa Rican conservation. The park contains half of all species found in the country and 5 percent of the planet's biodiversity, making it one of the most biologically diverse places left on Earth. Yet it occupies just 105,000 acres of land, plus 10,000 more in the ocean— tiny by global standards. (The threshold for qualifying globally as an "intact forest landscape" is 125,000 acres.) You could fit more than twenty Corcovados inside Yellowstone National Park.

Corcovado's A-list celebrity is the jaguar. As apex preda- tors, jaguars play a key role in regulating the whole system by keeping in check populations of prey animals that would other- wise overgraze and alter the ecological balance. But they need more space than the park provides. If your goal is to protect jaguars, then you also need high-quality jaguar food—which, here, includes herds of white-lipped peccaries. These hairy pig-like animals can travel six miles a day, in groups of up to 300, and they once roamed across large areas of Central and South America. White-lipped peccaries are also important as "ecosystem engineers": they spread the seeds of important food sources—by eating fruit and then pooping out the seeds some- where else, where they can take root and grow—and create watering holes, by rolling in the mud, that other animals rely on.

White-lipped peccaries on Costa Rica's Osa Peninsula. These pig-like critters once roamed across Central America but now exist in just 10 percent of their former range. They are an important source of food for jaguars and also play a key role dispersing the seeds of fruit trees. *Charlie Hamilton James/ NPL/Minden Pictures*

Today, white-lipped peccaries remain in just 10 percent of their original Central American range, thanks to deforestation and fragmentation as well as illegal hunting. (Far more common is their smaller cousin, the collared peccary.) Even those remaining populations are vanishing quickly, across an area from Mexico all the way to Panama. A report published in 2020 was titled *Precipitous decline of white-lipped peccary populations in Mesoamerica.*

In a study Whitworth facilitated a few years ago, researchers working with Osa Conservation set up 120 camera traps across an area of primary and secondary forest—so-called old-growth forests that have never been intensively logged, and others that were logged at some point and have since regrown—on the Osa Peninsula and the mainland across the Golfo Dulce, where there is another small national park called Piedras Blancas. They wanted to understand how land management in the region influenced which species were living where. They found, not surprisingly, that the greatest number of species lived in areas with the strictest land protections— the national parks—and also that the mix of species was significantly different in the landscapes humans had transformed. Large-bodied animals were missing from those "disturbed" areas—an absence that could have a big impact on the health of the whole system.

While the protected areas in general contained more species, though, the two parks weren't created equal. Jaguars, white-lipped peccaries, tapirs, and margays—a type of small, nocturnal wild cat—lived in Corcovado but not in Piedras Blancas. The boundaries of Piedras Blancas include the Golfo Dulce as well as a large river—two natural barriers—and also a highway. That geography, it turned out, made the park isolated and island-like. To protect jaguars and white-lipped peccaries, you need safe pathways between patches of habitat. You also need

to think about climate change and provide animals with routes to migrate to the habitat they will need.

This often means moving up in elevation. As the planet heats up, the conditions once found at sea level may now exist 1,000 feet up. It's not just about the actual temperature. The natural world is full of complex interactions, and as climate change alters ecosystems, even in seemingly tiny or impercep-tible ways, animals' existing home territories may no longer be adequate, and they will venture off in search of what they need. Whether they can reach those potential new habitats depends on the matrix—on what's in between.

Whitworth wants to protect "this elevational shift"—places, he said, where "if you build protected areas, or some kind of functional system, you're going to get as much wildlife biodi-versity through a climate bottleneck as possible." The key is finding sizable protected areas at different elevations, and then working to build a corridor between those places. It's a model, he says, "that you could pick up and put on any kind of area where you've got some kind of relatively intact habitats around elevation." A study that had just been published showed that during extreme drought in 2015 in Papua New Guinea, forest birds had moved up in elevation; the number of birds in the low-lands had declined by 60 percent, while the number of birds above 1,700 meters (about 5,500 feet) had increased by 40 per-cent. Whitworth's vision for the Osa Peninsula was something he called "ridge to reef," a way to reconnect ecosystems so the local wildlife could roam, from the sea to the mountains.

After a quick breakfast of huevos rancheros, we were headed up to the town of San Vito, partway between Corcovado and La Amistad, another national park. La Amistad lies in the Talamanca Mountains, a vast wilderness area and United Nations World Heritage Site on the border with Panama. The mountains contain one of the largest remaining forested areas

in Central America, home to 215 species of mammals, 600 species of birds, 250 species of reptiles and amphibians, 115 species of freshwater fishes, and 10,000 species of flowering plants.

The sweep of land from Corcovado to La Amistad, which Osa Conservation refers to as AmistOsa, spans roughly 200 kilometers (about 125 miles) and encompasses more than 1.3 million acres of land and sea. It contains a handful of designated biological corridors, and some animals already move through them, to varying degrees of success. In between the two parks, though, the matrix includes the Pan-American Highway and the Central Valley, a farming and ranching area containing cattle pastures, palm oil plantations, and pineapple farms, all privately owned by a mix of independent farmers and big conglomerates. Without a clear path uphill, the scores of species that call Corcovado home could be in trouble. Christopher Beirne, a data scientist who works with Whitworth, has built computer models to understand potential routes for animals needing to move between the two big protected areas. Using those as a guide, Osa has set out to plant trees across this landscape to build a wildlife corridor.

— – – —

Between 1940 and the late 1980s, Costa Rica lost nearly two-thirds of its forests, in what Whitworth has called "rampant deforestation associated with bananas, cows, and gold." But then the country pivoted, committing to protect nature and invest in ecotourism as an alternative to resource extraction. Over the next several decades, forests began to grow back; forest cover has increased 150 percent since the 1980s. More than

Although more than a quarter of Costa Rica's land is protected from development, much of the rest of it is no longer forested. Pineapple plantations such as these require vast amounts of pesticides and offer no resources for wildlife on the move.
Michael Pitts/NPL/Minden Pictures

28 percent of the country's land is now protected from development to some degree, with 10 percent set aside in national parks. As a result, nature here is doing far better—and its future is more promising—than in many other parts of the tropics. But like anywhere else, there are still threats. Conservation is a constant game of defense. Lack of enforcement of the environmental laws is an ever-present issue.

There are also the pineapples. Costa Rica grows over $1 billion in pineapples each year, nearly half of the total global pineapple trade. The crop requires multiple applications of more than a dozen different toxic chemicals—Costa Rica uses more pesticides per capita than anywhere else in the world—making pineapple farms vast ecological wastelands. And, of course, there is climate change—which is exposing the potentially catastrophic weak points in Costa Rica's otherwise prescient eco-planning.

Whitworth took the helm of Osa Conservation in 2018. Since its founding in the mid-1990s, its mission had been to protect the matrix outside Corcovado. In the 1990s, tropical biologist Adrian Forsyth was working for the nonprofit Conservation International (CI), and he attended a board meeting on the Osa Peninsula. The scientist in charge of CI's Costa Rica and Panama programs at the time, a Costa Rican named Manuel Ramirez, owned property nearby and invited some guests from the meeting for a visit. Ramirez's father had built a small shack down on the beach, and one day Forsyth left a note in the guest book there: "Manuel, I'm going to be jealous until we are neighbors." A few months later, Ramirez wrote to say that his neighbor was selling. "I convinced my wife that we needed a property on the Osa," Forsyth recalled, "and we bought it."

Forsyth was based in Washington, DC, and he envisioned that the Osa would simply be "a place to go and hang out." But the more time he spent there, the more it got under his skin.

He and Ramirez worried about the size of Corcovado relative to the density of wildlife living within its boundaries. This was becoming an increasing concern globally, as scientists began sounding the alarm about the inability of protected areas alone to support healthy populations of wildlife, and as humans continued to dismantle the lands surrounding them. Corcovado was critical to the survival of Costa Rica's biodiversity, but despite being the country's largest national park, it was far too small. The future of the species that lived there—jaguars, peccaries, pumas, tapirs, anteaters—depended on making the land *outside* the park's borders wildlife-friendly.

There were some small conservation groups operating nearby, but they kept shutting down for lack of funding. Forsyth and Ramirez mused about forming their own nonprofit, and when 300 acres of rainforest on the Osa went up for sale, they determined to buy it. Forsyth approached Gordon Moore, the tech legend and cofounder of the Gordon and Betty Moore Foundation, about funding the purchase. Moore gave him $300,000, and the organization Friends of the Osa was born. (The organization later changed its name to Osa Conservation to better reflect its mission.)

At the time, a growing body of research was making clear that linking up protected areas was critical. The Costa Rican government was talking about delineating official biological corridors. (It has since officially designated forty-four corridors throughout the country, calling these "the second most important conservation strategy in terms of territory and scope," but planning, regulation, and enforcement are spotty or nonexistent, and some of the corridors are little more than lines on a map.) A related law allowed landowners to form wildlife reserves that could shield their property from the long arms of municipal governments, which were pushing development and sometimes used eminent domain to slap down roads across

private properties. Forsyth and Ramirez got together with some nearby Osa neighbors, including the eco-resort where the board meeting had taken place, and formed a wildlife reserve, signing up their own parcels along with the 300 acres Moore had paid for. Today, that reserve has grown to include more than 7,000 acres packed with primary and secondary forest.

"We just knew from the fundamentals that a park the size of Corcovado is going to *leak species*," Forsyth recalled, a term ecologists use to mean that the wildlife inside the park would wander beyond its boundaries. If the park's wildlife thrived and its populations grew, some animals would necessarily seek out new habitat in the surrounding areas. "Which is what's happening now," Forsyth said.

We were sipping coffee one January afternoon on a hillside at Osa Conservation's campus. Forsyth's role these days is as an informal science advisor; he had recently left the board and even sold his property. ("I gave the money to Stanford," he likes to joke, referring to college tuition for one of his three sons.) But he still visits. On this particular occasion, he had come to study moths. He said he was trying to learn to perceive a moth the way a monkey or a bird might, rather than the way humans do. As we watched a procession of iridescent hummingbirds visit the flowers just beneath where we sat, Forsyth talked about climate corridors.

In the 1980s, while studying insects in the Peruvian Amazon, he had been gripped by the notion that climate change would ultimately force animals to migrate uphill. He recalled traveling down the eastern slope of the Andes, heading into the Amazon, and thinking about how, as climate change drove species uphill, "these elevational corridors were the short-term solution, where things could find their climate in a small amount of space." But if you were a creature living "out in the flatlands of the eastern Amazon in Brazil," Forsyth said, "you might have

to go 1,000 or 2,000 kilometers to re-find your climate. And oh, by the way, it's a soybean field." That region lacks what scientists call "climate resilience." But where the Amazon meets the Andes, and where ecosystems are still intact, "things can go up a couple hundred meters and they're in their comfort zone." A lot of Forsyth's research over the decades had been about the question of how fast things have to move. "Dung beetles have to go up forty meters a year to keep pace with the climate to which they're adapted. That's what's happening, and we know that because we have twenty years of data of things moving up the mountain."

All the way back in 2003, research showed that species around the planet were shifting their ranges north or south—toward the poles—by an average of just over six kilometers a decade. A 2011 study found that species of both animals and plants had been moving up in elevation for a century, by an average of thirty-six feet every decade. More recent research has found that tropical species seem to be moving uphill between two and ten times more quickly than their temperate counterparts. In 2018, scientists discovered that on one Peruvian mountain, birds that had once resided at the highest elevations had already disappeared, and other birds were living in smaller populations and smaller areas and were at risk of extinction because they had nowhere higher to go. The researchers called it "an escalator to extinction."

— — — —

San Vito sits in the southern Pacific highlands, less than ten miles from Panama. The rugged landscape hosts 800 species of butterflies and 60 species of bats, among hundreds of other animals and thousands of plants. But the forests here are severely fragmented, with small remnant forest patches interspersed with farm fields and cattle pastures. Corcovado and La Amistad constitute a single "natural biological unit," as Forsyth put it.

If animals from the Osa are going to reach the Talamancas, they need to traverse this area. Without that connectivity, he said, the animals that can't move out of Corcovado "will hit the wall." So Osa Conservation is working with local landowners to regrow forest corridors here.

Tree planting has become something of a cliché, maligned as a way for people in the US and Europe to assuage a seeping guilt around overconsumption and inaction while continuing to live a comfortable, carbon-intensive lifestyle. In fact, many efforts in reforestation have been huge failures, done without buy-in from local people or a sense of a landscape's history, or an understanding of what species are important. (There have even been proposals for extensive tree-planting in African grasslands that are mistakenly viewed as deforested when in fact they are ancient savanna ecosystems, "landscapes supporting herds of megafauna and livelihoods for hundreds of millions of people," as several researchers wrote.)

Often the tree-planting projects are not unlike the paper corridors, lip service without substance. Researchers who examined reforested areas near San Vito found that half the forest had been cut down again within twenty years, and 85 percent within half a century. If reforestation isn't done right, said Whitworth, then the land "becomes recleared very, very quickly." Programs that offer people a chance to plant a tree for a dollar or so don't account for the long-term maintenance or any cost to the landowners, so they often fail. Whitworth partnered with a local economist and a student to explore the true

The common blue morpho butterfly (*Morpho helenor*) is an official national symbol of Costa Rica. Its range extends from Mexico all the way to Argentina. Costa Rica is home to 1,200 different butterfly species, delighting tourists who come to the country for its rich biodiversity. As the climate shifts, species of all kinds will have to move to higher elevations to meet their daily needs. *Piotr Naskrecki/Minden Pictures*

cost of planting trees. "We asked the question, What's the cost to see one of our trees planted, a tree in thirty years' time, guaranteed?" The real cost in Costa Rica, it turned out, was fifteen dollars. For their reforestation projects here in the highlands, Osa Conservation pays for everything, so there is no cost to the landowners—only potential benefits.

Rodrigo DeSousa, who oversees the project, lives in San Vito and sits down with property owners to build specific tree-planting plans based on their needs. "I feel many of these trees won't be harvested, as I think the value of standing trees will go up from here to the future," he said. "I mean, if you came here twenty years ago, nobody cared about trees." DeSousa, who is Venezuelan, ran his own landscaping business there before leaving when Hugo Chávez came to power. He holds master's degrees in agricultural and resource economics from the University of California, Davis, and in environment, development, and peace from Costa Rica's University for Peace. This combined background seems to make him uniquely suited for the task of building a network of landowners eager or at least willing to have more trees. When I visited, DeSousa had recruited nearly 300 farms for the network. One farmer was giving up raising cattle and wanted to build a series of cabins in the woods—the more woods, the better. Another landowner was a local lawyer who had recently purchased a twenty-acre property for the sweeping views. She was thrilled that someone might plant and maintain a patch of forest.

But that maintenance is key; you can't just stick seedlings in the ground and leave. DeSousa and his team ask landowners what benefits they most want—some shade in a cattle pasture, improved soil productivity, or perhaps even a wholesale switch from agriculture to ecotourism. Then they determine the right assortment of trees, starting with fast-growing types like balsa and legume species that help put nitrogen in the soil for other

plants to use. They choose trees targeted toward specific animal species—like oak for agouti, or trees in the avocado family for birds. They also think about a mix that will help spark further regrowth. "We want natural regeneration," DeSousa said. "When trees grow, birds come in, and then they spread the seeds." The team also plants trees that may be of particular interest to the farmer or landowner, because "it's not just about biology, it's about people and relationships." And they plant rare trees, a few in each area.

To fuel this restoration, Osa Conservation runs five greenhouses, each located at a different elevation and tailored to the trees that live in specific types of forest. At any one time, as many as 125,000 seedlings are growing in the greenhouses. "There is no recipe for restoration," Jose Rojas, the group's lowlands restoration coordinator, told me. "In every forest, every country, it is different." In 2017, the organization started an experiment to kickstart forest restoration on a portion of its own land that had been cleared decades ago for cattle pasture. On a series of forty plots across fifty acres, they planted varying percentages of balsa trees—the species known for the lightweight wood used in model airplanes and other crafts. Balsa is a "pioneer species," a tree that quickly creates a canopy that can shade out aggressive grasses and vines. They also planted sixty other species of native trees. The idea was to determine how to create the healthiest forest regeneration. It will be several more years at least before there is clear data. Some of the plots also contain "rewilding" features—bat houses and bird boxes meant to attract animals that can help accelerate the reforestation process by pollinating plants and dispersing seeds. "It's a very slow process," says Rojas, who is from San José, the capital. "I may not see the forests, but I do what I can while I can."

Even all this work is only part of the process. Once you've convinced the owner and selected the trees, generally about 400

per acre, you have to transport them there, sometimes using a tractor or horses, place them lovingly in the ground, and tend to them. You have to prune back the grasses that might otherwise out-compete a little tree until a canopy establishes, and build fences to keep cattle out, and check on the trees to make sure they are safe from disease. DeSousa's restoration field crew returns several times a year to maintain the young trees, which also helps maintain the relationship with the landowner. Then you have to wait for the trees to mature enough to fruit, and for animals to come and distribute the seeds so more trees can grow, and then you have to wait several decades or more for the land to begin to approximate a forest again, though it may never again have the same assemblage of species it had to begin with.

Humans are great at building things—highways, tunnels, dams, skyscrapers, cities—and we can send spacecraft to other planets and the moon. It seems like it should be fairly easy to build some simple links between patches of forest. But it really isn't. Regrowing a rainforest is something that requires no concrete, no engineering feats, no complicated machinery, no government permits. It requires, instead, ecological knowledge, local knowledge, passion, goodwill, and dedication, plus painstaking manual labor. These may all seem simple and relatively low-cost, but trying to reconnect broken landscapes is a daunting effort that also requires a great deal of time. Trees take years, decades, centuries to grow. On the other hand, given the right conditions, one person can burn down or bulldoze a chunk of rainforest in no time at all.

The restoration crew from Osa Conservation at work on a replanting project near San Vito, in Costa Rica's Pacific highlands. Tree-planting efforts can often be failures, so the Osa team collaborates with each landowner to ensure the program meets their needs. They also maintain the seedlings, select a balanced mix of trees for specific animal species, and plan each project to spark natural regeneration of the forest. *Ian Rock/Osa Conservation*

— – – —

It was the rainy season in Costa Rica, the time when planting takes place. A group of about a dozen workers from the restoration crew was preparing to plant trees on a former cattle pasture. We stood on a little mesa as another Osa staffer, Rodrigo Benavides, the highlands restoration coordinator, launched a drone so we could see what the topography looked like. Not far away, the land dropped steeply—good habitat for trees but not for cows. Benavides also showed us satellite imagery of the area and pointed out the paths they were trying to regrow as forested avenues. DeSousa gestured across the landscape, to the property line. "We're gonna plant almost to the border of the farm," he said, and once that was done, he hoped to convince the owner to reforest even more. "Sometimes what we do is, we do something this year, and then we develop that trust with people, then we can come back and say, 'Hey, this could be important, to do like another strip.'" Osa Conservation's goal is to plant 700,000 trees by 2027.

"It's a big job," Whitworth admitted. "But if you look back in time, right? You've seen this focus from conservationists back since the '60s. And you see that with hard work, things have generally gotten better down here on the peninsula side. I think if you look at this landscape in thirty years, it could be very different."

— – – —

Back down toward sea level, on the neck of the Osa Peninsula between mangrove wetlands to the north and the national park to the south, nestled among forest that provides a path for animals moving toward the highlands, sits the village of Rancho Quemado. It was settled by miners and loggers who arrived on the Osa from Panama and other parts of Costa Rica in the mid-twentieth century to extract its gold and timber. Each year, white-lipped peccary herds migrate in and out of Corcovado

and pass right through the village—where the residents used to hunt them in unsustainable numbers. A few decades ago, the herd numbers were crashing. But today, Rancho Quemado functions as the ultimate protector of the peccaries.

In 2012, a new generation of residents partnered with nonprofits and universities to create business opportunities centered around conservation. On the way back from San Vito, Whitworth and I paid a visit. I stayed in a still-under-construction bed and breakfast, whose owners hoped to benefit from their proximity to Corcovado and their thriving forests that served as a buffer zone for the park. Yolanda Rodriguez, who was building the B&B with her husband, said she had taken the "passion and vision" from her father's generation, ex-hunters who had wanted to start a youth group to teach kids to value the wildlife but never managed to do it. Rodriguez and some of her neighbors led us on a hike through the forest, but not before everyone picked some rambutans, a fuchsia fruit covered in short tentacles that make it look like some kind of tiny alien life-form. The rubber-boot-clad group filled pockets and bags with fruit and walked the dirt road to the edge of the village.

It was late in the day, and we followed the community trails into the forest, where we hiked beneath scurrying monkeys and chattering birds as dusk descended. On a wide path that looped back toward the village, our feet squished in the soupy mud and dozens of frogs and toads hopped out of our way. Concerned that I'd inadvertently squish one, I tried to make them out ahead of time. But they were nearly impossible to see until they were flying past my boots. A tiny deer stood frozen for a moment on the trail ahead of us, before trotting off into the trees. Finally, we emerged at a clearing with a large, raised platform, almost like a stage for some kind of community ritual. It was a viewing platform for watching the peccaries. Rancho Quemado's

residents now keep tabs on the herds and even occasionally help with radio-collaring. The community also hosts an annual peccary festival—an event that used to involve killing some of the animals for food, but now celebrates their safe return and further protection.

To Whitworth, the transformation of San Vito from peccary eaters to peccary benefactors is a model of success. It wasn't imposed on anyone; the community spearheaded it. "Conservation has to be about working with communities, with people—the whole concept of coexistence," he said. Coexistence was on the minds of ecologists and conservation professionals everywhere I went to report on connectivity. The specifics were different, but the larger theme was the same. We can't partition the globe and say, "Wildlife here, humans there." We can't expect wild animals to observe boundaries humans have drawn on a map. We can't dismiss the concerns of communities who live with the threat of predators picking off their livestock, devouring their crops, or injuring their families. We need to find new ways to live with wildlife—something humans did for thousands of years. After all, the whole world used to be the matrix.

— — — —

After a few days of traveling through Osa Conservation's projects, including mangrove restoration and a monitoring project on the Golfo Dulce, we headed to Piro, the group's campus in the forest. From my room on a rise at the field station, I could hear the ocean, far below. A toucan called out from a branch nearby and spider monkeys leapt through the trees just past my balcony. I could have happily sat there all day, but the forest

A spider monkey in the canopy on Costa Rica's Osa Peninsula. The Geoffroy's or black-handed spider monkey, which ranges across Central America, is listed as endangered due to habitat loss and other threats. *Luca Eberle*

was calling. Whitworth and Charapita and I headed up through a section of forest filled with massive fig trees. It had been logged around seventy or eighty years ago, and the trees here were only forty years old. The forest had come back on its own.

The land had been logged for agriculture and then abandoned, Whitworth said. "It's just natural regeneration, no action plan. Which, to be honest, works really well if you're next to a primary forest. If you're next to a place where there are fruit-dispersing animals that can get there quickly, natural regeneration works really well." For his PhD research, Whitworth had studied the conservation value of secondary forests. "It's context dependent—on how far you are from old growth, how far you are from animal populations, how long the land use occurred. I call these forests, where you've got a lot of old growth close by and the soil isn't heavily degraded, a best-case scenario."

To our right, a bit farther up the trail, Whitworth pointed out an ajo tree, named for its garlic scent. It grows huge and then hollows out, providing a home for bats. To the left, he pointed out the spot where they would soon install an observation tower, nearly one hundred feet high, and a bridge over the canopy, where staff and visitors could climb for a monkey's-eye view.

As we walked through the forest, Whitworth kept pointing things out, often stopping in the middle of a sentence to show me something new: baby anole lizards, a great tinamou, a golden orb weaver spider, tapir tracks. We talked about the risks from new development on the peninsula, where wealthy Americans were building fancy getaways. Driving the road to Piro, we'd passed a giant mansion built by a Californian who illegally cut down seventy trees and then built a fence that looked to be more than ten feet high, forming a barrier to wildlife movement.

"I think part of the challenge with biological corridors is that landowners don't even know that they're buying in a corridor,"

Whitworth said. He wondered if he should be working with real-tors, trying to talk to them "about making people aware about where they're buying and what that means and what it could do. Maybe that's the pragmatic approach. You can very—" He inter-rupted himself. "Oh, this is an amazing tree!" He pointed to a huge tree with giant buttress roots, *Tachigali versicolor*, known as the suicide tree. They grow quickly, reach perhaps ninety years old, and then flower once, en masse. Then they die. "Every seven, eight, nine years, a bunch of them will go together. That kind of lets the forest flow with the seeds. Then they just topple over, and one of the youngsters is going to fill that light gap."

We reached the old-growth forest, a portion that had only ever been selectively logged—a tree here, a tree there. Whitworth pointed to the stilt palms, whose strange, above-ground roots form a teepee-like structure. A third of tropical rainforest trees are palms, but most tree-planting programs don't include them; they don't consider palms to be trees, because they aren't hardwoods. "They're really valuable food resources for wildlife because the fruits of palms are usually quite oily. High fat content. So if you're building corridors with tree-planting programs, and not considering palms' incredible diversity, you're not really growing true, neotropical rainfor-est," Whitworth said. "That has to change if the goal is to create corridors to benefit biodiversity."

Farther up the trail, he pointed to a hole high up in a tree trunk. He said he'd seen scarlet macaws going in and out. "I think they're trying to decide where to nest." A rope hung from a branch, 150 feet up from the forest floor. Whitworth had put it there when he'd climbed up recently to set a camera trap, to watch the macaws but also in the hope of seeing something rarer: a harpy eagle. These large, hook-beaked predators once lived throughout Costa Rica, but deforestation wiped out the population. Someone had recently sent Whitworth a recording

of what sounded like a juvenile harpy, from right near where we stood. And just a few weeks earlier, someone had photographed a harpy up north, causing a national stir at the possibility of the birds' return. Whitworth was increasingly convinced that they were back on the Osa, too.

He'd been hiking one day and stumbled on a spider monkey skull half-buried in the leaf litter, with four holes—three in the top and one at the base—that were clearly made by talons. "An eagle had obviously killed it instantly and eaten it," he said. "But there's only two eagles that can do that." It had to be either a harpy or a crested eagle, both rare. Whitworth had hatched a "kind of nuts" idea about how he might find a harpy, if any were indeed around. "I'm thinking of nailing some stuffed sloth teddies to some of the branches, just to see if anything like a bird of prey comes in," he said. "Everybody thinks I'm bonkers, but I'm convinced it'll work." He was waiting until the dry season so the stuffed sloths wouldn't "end up really manky." "People have put little rubber snakes out on the trail with a camera trap, and little falcons and stuff will come in, so I don't see why a big bird of prey wouldn't go check it out," he said.

We paused at a tamarind tree. Spider monkeys prize its fruit, and that's how the tree gets the monkeys to move its seeds around. The seeds have evolved so they need to go through the monkey's gut in order to germinate. "A lot of these trees can't exist without the wildlife," Whitworth said. "If the corridors are empty forests, then they're not functional."

Biologists are increasingly worried about this idea of empty forests; they talk about the phenomenon of "defaunation," a diminishing of the creatures that once called a particular landscape home. Extinction is incredibly consequential, but it is just the extreme end of a continuum. (Around 322 vertebrate species have gone extinct in the past 500 years.) Defaunation, on the other hand, is a broader idea. It refers to both the vanishing

of a whole species and also the loss of specific populations or just sheer numbers of individual organisms. Even though a species might still exist, its numbers could decline so dramatically that it can't perform its ecological role—spreading the seeds of a fruit tree; creating potholes that fill with water, where other creatures can lay eggs or drink; fertilizing the soil through poop. Defaunation can happen quietly; you might not notice that there are fewer birds singing in your local woods than there were a few years ago, until one day the forest is silent.

A Brazilian ecologist named Alberto Campos, who works in a growing field of science called rewilding, echoed Whitworth's concern about empty forests when I spoke to him via Zoom. "A lot of the trees have developed some kind of interaction with animals to expand their ranges or reproduce," Campos told me, "so I'm worried about these cascades. When you remove the faunal element, what happens to the trees?" Rewilding is one way to keep ecological systems functioning. The researchers of the collapsing food webs study concluded that "natural recolonization and reintroduction of native mammals to their historic ranges" could help reinstate the missing links in the webs.

This was something Whitworth was thinking seriously about, too. When I visited Osa a second time, he was building a rewilding program. He'd hired a new employee, a Costa Rican biologist who had worked for the government and understood the inner workings. Whitworth hoped to reintroduce harpy eagles, but that "could be contentious and difficult." So they were working on something that was more of a stepping stone, a less controversial effort that could prime people to the idea of species reintroductions. Whitworth hoped to move some white-lipped peccaries from outside Corcovado over to Piedras Blancas. "They just redid the management plan for that park, and they are open to some of these proactive steps," he said. After that, he wanted to move to giant anteaters, another locally extinct animal.

In the meantime, he had also established a research group to study movement ecology, a new scientific discipline that uses data from technology like radio collars and satellite tracking devices to provide insight into animals' migration routes and use of all sorts of landscapes throughout their lives. The group was putting radio collars on a variety of animals, including king vultures—which play a crucial role in ecological connectivity by recycling nutrients and keeping parasites at bay. Old-world vultures—those in Africa and Europe—were experiencing "some of the most spectacular declines among vertebrates," said Beirne, with fourteen out of sixteen species listed as threatened or near-threatened, largely due to eating poisoned carcasses. Conservationists were worried about the same thing happening in the New World—the Americas—where two out of seven vulture species were listed as threatened. "We know very little about these vultures," Beirne said. "We have to act now." They had mapped the movement of twelve king vultures fitted with "backpack" transmitters and found that all had at some point left the Osa Peninsula, but none, at any point, flew over the ocean. They almost always flew over forested areas. It was just the bare minimum of information, but it was a starting point for understanding what the vultures needed and how habitat fragmentation might be affecting them.

Beirne had also expanded his wildlife movement models to cover not just AmistOsa, or even Costa Rica, but all of Central America. The models use various methods for finding the best

---

This king vulture, which the Osa Conservation team fitted with a backpack transmitter, will provide them with critical data about the behavior of this key species. Vultures play an important ecosystem role by recycling nutrients—providing a form of ecological connectivity that helps keep the whole system functioning. _Luca Eberle_

corridors, including a tool that sends electrical currents on the least resistant path from A to Z. The work involves assembling layers of data, things like a "human modification index" that shows land-use change, and a "resistance surface," which is based on which area a given species will or won't use. "What we're saying is, within each country there are regions where you can build climate conservation," Whitworth told me. He called the effort "a roadmap to save biodiversity in our changing landscape." Not unlike Gonzalez's maps for Montreal, the idea was to show where limited conservation money might best be invested. They released a report, called *The 10 Climate Adaptation Lifeboats Across Central America*, that they hoped could guide a conservation agenda across the region.

— – – —

One afternoon, Eleanor Flatt, an English wildlife researcher who does a bit of everything at the field station, opened an app on her phone to show me the path of an ocelot that had been released the previous week, after it had been hit by a car elsewhere in Costa Rica and then rehabilitated. Rather than stick around near or inside Corcovado, the cat had made a beeline for the coast. It now seemed to be wandering north along the beach. It might be, said Flatt, that there were already too many ocelots with territory nearby, and this one was in search of less-populated terrain. Or perhaps it was headed back where it had come from. The routes of animals can be unpredictable—which can make corridor planning even more challenging.

After breakfast one morning, I set off into the forest with Flatt and some coworkers, who were climbing trees to swap out the batteries on camera traps set high in the canopy at either end of a rope bridge that the team was monkey-testing. Arboreal bridges, as they are called, are simple rope structures that can connect forest patches fragmented by roads. Animals that live in the canopy can easily cross these bridges, but researchers

have found that the particular design impacts which species will use it. They were testing three designs around Piro; one bridge crossed the road to the station from town. The researchers lured monkeys to the bridges by setting out tampons soaked in vanilla, with the idea that some animals might curiously cross, and then others would follow the example. Sometimes restoring connectivity was as simple as stringing ropes across a road.

Sometimes, though, it was far more complex, as the climate corridor effort revealed. As climate change and land-use change continue to interact and remake our world, we need innovative, all-hands-on-deck efforts that consider the impact people have on wildlife, while also recognizing that the presence of wildlife impacts people—for better and for worse. There are benefits and hazards to having jaguars cross your farm or village. So how can we create opportunities where both people and nature prosper?

I'd been researching different models of "community-based conservation" around the globe, and I wanted to see how these programs could, when done right, offer that kind of win-win solution. I found a perfect opportunity about 2,000 miles to the northwest, in the Mexican desert.

CHAPTER 4   *Central & Northern Mexico*

Entering the Agave Corridor: Connecting Bats, Plants, and People

A week before Halloween, when bats were appearing everywhere in my town—plastic ones hanging from doorways, paper cutouts adorning windows, cartoons on seemingly every supermarket label—I went to visit a place where bats were disappearing. I flew to Monterrey, Mexico, just an hour-long hop from Houston, and drove west through the morning smog that hung beneath the dramatic peaks of the Sierra Madre Oriental. Yucca-filled plains stretched toward the mountains, occasionally punctuated by a giant steel refinery or other industrial installation. Monarch butterflies fluttered across the highway on their southern migration. From the back seat of a Toyota Hilux pickup, I wished each one safe travels and hoped their journeys didn't end on a windshield.

Previous Spread: Biologist Ana Ibarra of Bat Conservation International and local community member Don Chalío set up equipment to monitor Mexican long-nosed bats at a roost in Sierra La Mojonera, a protected area in northeastern Mexico. These nectar-feeding bats were listed as endangered in the late 1980s, and since then their numbers have continued to decline. *Ruben Galicia/Bat Conservation International*

On the way out of town, I'd noticed a surprising number of Tesla advertisements around, and I soon discovered the reason: Elon Musk planned to build a huge car factory here, along with worker housing and other satellite development—despite the fact that this arid region of Mexico had, the previous year, suffered a water shortage so severe that the taps ran dry. The government had to deliver water each day to 400 neighborhoods, and small businesses were forced to shutter without water to function. Monterrey is Mexico's industrial capital, and a law lets manufacturing facilities draw water from the area's dwindling aquifers even as residents parch. (It takes about 725 gallons of water to manufacture one Tesla.) Topo Chico, the drink company, is also based here, and though the drought caused some fleeting shutdowns, the company was largely able to bottle water for export.

I had not come to learn about steel or cars or fizzy water or even monarch butterflies, but I was interested in the drought. It had eased up a bit in Monterrey since the previous year, but it was part of a drying trend that has changed rhythms of life across the whole region—including for a group of farmers whose lands were vital to the struggling migratory bat I had flown across the border to learn about. Southwest of Monterrey, four hours by car without traffic, lies a broad, flat, isolated desert valley that is home to dozens of ejidos, agricultural communities that share common title to their lands. A few of the ejidos had recently been abandoned, their residents forced to leave because there was simply no more available water. I was headed to visit some of the communities where people still eked out a living. I'd been invited along by two bat biologists—one American, one Mexican—who are working with a local conservation area and the ejidatarios, as the residents are called, to preserve and regrow a corridor of agaves that can help both bats and humans.

Community-based conservation—protecting nature by working in partnership with the people who live on the land and know it best—emerged several decades ago as a response to top-down environmental policies that strove to protect a myth of nature as pristine and human-free. Those policies also often pushed people off their traditional lands. But the community conservation model has grown increasingly important alongside a recognition that humans and nature can't exist in isolation from each other. Nature and culture are inextricably connected. "By restoring the habitat for the bats, we also restore the land for the people," Kristen Lear, the American biologist, had told me the first time we spoke.

I was driving with Lear and Ana Ibarra, the Mexican biologist, to Matehuala, a small city in San Luis Potosí. Lear, who lives in Colorado, wore tiny stud earrings in the shape of bats and carried a bat-patterned tote bag. She likes to say she got her unofficial start as a bat biologist at age twelve, in Ohio, when she built bat houses to earn a Girl Scout badge. A lifetime member of the Girl Scouts, Lear is the person you want to travel with; whether you need duct tape or hand sanitizer or toilet paper, she has it. "I've really loved bats since I was little," she said. She also loved birds, and in her first year of college, she worked as a field assistant for a grad student's research on cliff swallows. But the summer after her second year, she went to Texas to work on a bat project. "That's when I knew, you know? You fall in love with it." For her PhD research, Lear, who is now the agave restoration program manager for Bat Conservation International (BCI), lived in some ejidos, studying how both the communities and the migrating nectar bats used the agaves— and looking for ways to incorporate conservation into traditional rural livelihoods.

Ibarra is BCI's strategic advisor for endangered species, based in the southern state of Chiapas, near Guatemala. She,

too, began her science career studying birds—but she likes to stress that she was "an ecologist that works with birds, not a hard-core ornithologist. They are like a species of their own." Ibarra studied the impacts of habitat fragmentation on tropical forest birds. Then she became interested in learning whether birds or bats were better pollinators—"As every biologist will tell you, for every question, it depends"—and as a postdoctoral researcher, she began her drift bat-ward. On our trip she wore hummingbird earrings, but she swore that "bats are way cooler."

We met up for dinner in Matehuala with Lissette Leyequien, who works for CONANP (Comisión Nacional de Áreas Naturales Protegidas), the federal government agency that oversees protected areas. Leyequien is director of a small conservation area called Sierra La Mojonera, which lies in the remote mountains near the border between San Luis Potosí and the neighboring state of Zacatecas. It's made up of land owned by seven ejidos, so the communities are essential to the reserve's connectivity with the larger landscape. Mojonera and the ejidos together occupy a landscape of agave habitat that is vital to the survival of nectar-feeding bats.

We had barely sat down when Leyequien said she couldn't contain the good news she was sitting on, even though technically she wasn't supposed to share it yet: The government was designating a new protected area adjacent to Mojonera, adding more than 770 square miles (200,000 hectares) of land. Lear and Ibarra had described Leyequien to me as "a mother hen," and it was immediately clear how much she cared not just about the lands and wildlife in her protection, but also about the people. "A few generations ago, people had lots of cattle, and it was easy to have them," Leyequien said. Back then, residents collected rainwater for their crops and livestock, and many ejidos had working wells. But now the wells were toxic, contaminated by heavy metals. The rains came far less frequently. Most

people raised fewer animals. Communities had to purchase water, which was trucked in from the nearest town. "Some people are not even trying now," Leyequien said. "They sell their cattle, don't even try to grow crops. They just survive."

In striking contrast to the way many rural US communities view representatives of federal government agencies, the ejidatarios welcome Leyequien as family, with hugs and meals. It's in part due to her particular demeanor, a potent mix of compassion, competence, and no-bullshit pragmatism; she's like a smiling steamroller you definitely want on your side. But it's also due to the fact that Mexico's protected natural areas are intertwined with people. In order to safeguard Mojonera's lands and wildlife, Leyequien knows she must protect the adjacent lands as well, and that means taking care of the families that live and work there.

— — — —

Three species of nectar bats move through this corridor on a binational migration between Central Mexico and the southwestern US. The one in the most dire straits is the Mexican long-nosed bat (*Leptonycteris nivalis*). Relatively large, with a fourteen-inch wingspan, this bat dines entirely on agave nectar for part of the year. It has ash-colored fur, a leaf-like nose flap protruding skyward from its snout, and an amazing, narrow tongue that can extend out three inches—about as long as its entire body—to get to the nectar. It lives mainly in desert-scrub woodlands and migrates along a 700-mile stretch from central Mexico up to southwestern Texas and the very bottom edge of New Mexico. Listed as endangered in 1988, Mexican long-nosed bats have only declined in number since then. In the 1990s, there were roughly thirty known roosting caves for the species,

Mexican long-nosed bats rely on agaves for food as they migrate between Mexico and the southwestern US. But land-use change has severely disrupted the availability of agaves along the route. ©*MerlinTuttle.org*

each containing thousands of bats. Today just ten of those sites remain, and scientists fear there are only between 6,000 and 10,000 of the bats left at all.

Ibarra spent nearly a decade working for a local nonprofit in Mexico, monitoring the single cave where Mexican long-nosed bats mate. This sole breeding site sits in an area northeast of Mexico City that is rapidly developing, with wealthy people from the city building weekend homes. The cave, called Cueva del Diablo, was increasingly under threat—from visitors wanting to explore it, from a road directly above it that was once quiet but suddenly rumbled with trucks, from proposed new construction dangerously close to the cave, where jackhammering could cause a collapse. Ibarra had grown frustrated that her job involved study-ing the bats' decline but "not intensely doing conservation," she said. When the chance came to work with BCI, she took it. Part of her job now involves working with the community around Cueva del Diablo—people she knew from years of monitoring the bats there—to push for smart development that won't destroy the mating roost.

Mexican long-nosed bats mate in the spring, and then the pregnant females head north, surfing a wave of agave blooms up through northeastern Mexico. Along the way, they stay in caves or abandoned mines for days or weeks, until finally reaching their maternity roosts on either side of the border. There, these endurance-athlete moms give birth to one baby each, and those little bats must stay in their caves for six weeks before they can fly. In late summer, the moms and their young make the journey in reverse, feeding on species of late-blooming agave.

When agaves bloom—which they do just once in their lifetime, sometimes after a decade or more—their stalks erupt in a giant array of nectar-filled flowers. Several species of migratory bats depend on this nectar as their primary food source. *Chris Gallaway/Bat Conservation International*

Agaves flower just once in their lifetime—taking ten to fifteen years to do so. They grow a giant stalk that towers as high as twenty feet, depending on the species, and then erupts in a feast of nectar-laden flowers. Afterward, they die. Their dried-up stalks remain for a time, like ghost trees rising from the scrub. Agaves were once plentiful across the bats' entire migration route. But now, the bats are finding fewer and fewer of the flowering succulents to fuel their journey. There are a number of reasons for the agaves' decline. Urbanization and development, along with agriculture and livestock grazing, have destroyed large swaths of habitat, leaving less and less undisturbed desert scrubland. The prolonged drought has made agave survival and regeneration more difficult in the areas that remain. Global warming has left this region, like so many others around the world, hotter and drier. When excessive heat and low rainfall reduce the amount of grasses available for cows and goats to eat, the hungry animals turn to other plants, including agaves. They will eat both the leaves and the stalks, which are sugary and rich in energy. Ranchers will sometimes cut the stalks down at the base just as they are flowering and leave them in the field for their herds. This deals a double blow: no nectar for bats and no pollinated seeds to grow into new agaves.

All of this causes a feedback loop. The destruction of landscapes once filled with native plants leads to further desertification. Native plants hold water that keeps the ecosystem lusher; when people clear the land for farming, or allow livestock to overgraze it, the entire system gets drier. The many gold, silver, and copper mines in the area further dehydrate the system by

Mexican long-nosed bats, which were listed as endangered in 1988, are under constant threat. Today there are just ten known roosting caves for the species and only one where the bats are known to breed. *Jon Flanders/ Bat Conservation International*

sucking up groundwater, making it more difficult for agaves and other plants to grow.

Unlike animals that walk, hop, crawl, slither, or otherwise move along the ground, bats fly through the air, so it might seem strange that they even need a corridor. But even flying creatures need access to suitable food and habitat. Mexican long-nosed bats, as well as their cousins the lesser long-nosed bats, which also migrate through here, evolved with a pathway of blooming agaves—a nectar highway of sorts. (Lesser long-nosed bats, whose populations have recovered in recent years, also feed mainly on agave nectar for most of their journey— along a more westerly path—but at the northern end of their route they eat nectar from large flowering cacti like saguaro. This varied diet likely helped their recovery.) Nectar-feeding bats, like guests at roadside hotels on asphalt highways, also need warm, safe places to sleep. But over the last twenty-five years, many of the original caves and abandoned mines that once sheltered Mexican long-nosed bats have become inaccessible to them. "They're empty of bats," said Ibarra. "They have been built over, they have been closed off, they have been used as dumpsters." Constant disturbance and even sometimes burning of trash keeps the bats away. "At least twenty of those roosts no longer host bat colonies."

Bats in general are struggling, all over the world. About a quarter of all bat species are either endangered or threatened, and in the US, more than half of all bat species are "at risk of severe population decline in the next fifteen years," according to BCI. "That's a lot of species that are under threat," said Lear. North of the Mexico–United States border, much of the land where the nectar bats feed is owned by the Bureau of Land Management and leased to ranchers for grazing, which has destroyed agave habitat. More recently, wind energy has become a problem, uprooting native plants and also posing a

mortal danger for bats that get caught in the turbines. (Wind turbines are a problem for nearly all bats, and could soon lead to spiraling decline even in species that are currently doing fine.) Mining is also an issue in the US, so BCI works with mining companies to replant agaves in adjacent areas.

Tequila and mezcal production, paradoxically, also threaten agave-feeding bats. Producers often clear native plants to put in a monoculture crop, then harvest the agaves before they flower, leaving nothing for bats to eat. Some smaller producers, mainly of mezcal, let a portion of the plants flower. But most don't. A bat-friendly tequila label aimed to certify farms that let at least 5 percent of their crop flower, but only one brand was using the label when I last checked. Consumers, Ibarra pointed out, could demand change. "People that think they are totally removed from the game have the power to change the game," she said. "They're not removed from this. There's responsible consumption."

To help save the nectar-feeding bats, BCI is working to replant and conserve agaves all along the bat migration corridors, and to safeguard their roosting sites. This means partnering with landowners to restore degraded lands, harvest seeds and grow new agaves, and even monitor bats. "The idea is that if we can restore the soil, water, native grass, that will create healthy habitat for these agaves to flourish," said Lear. All of that helps the ejidos, too. These farming and ranching communities also "need healthy, resilient land to grow their crops."

Because agaves take so many years to flower, the work is not just about bat-plant and bat-people connections. It's also about "connecting timescales," as Lear put it—planning for agaves, bats, and people to coexist many years into the future. Working through Mexican partners like CONANP and local nonprofits, the project provides training on regenerative agriculture and ranching, and help in developing sustainable businesses that support native habitat—rebuilding a chain of migration and

survival out of spiky desert succulents, using bonds made of people. "Bats are migrating such long distances, across boundaries, bringing people together," Lear said. "I always think about the social connectivity. They connect actors across that corridor."

— – – —

The Mexican long-nosed bats had made their late-summer journey south by the time I flew south myself to Mexico. The mother and baby bats had by this time moved well past Matehuala, and reunited with the males in caves near Mexico City, about four hundred miles south, where they would spend the winter. We spent the night in Matehuala, the nearest town to the ejidos, and in the morning we picked up provisions at the local Walmart and set out for the desert with Leyequien and two members of her staff. We crossed through vast areas of *izotal*, forests of the cartoonish yucca trees known in Mexico as *izote*—some single-trunked and topped by a head of spiky leaves, others with gnarled branches reminiscent of Joshua trees, but shaggier. The pavement ended and we bounced along a dirt road, across an increasingly remote landscape. Eventually we arrived at Las Huertas, a community of just a few dozen people in the state of Zacatecas, where María Isabel García Galván greeted us, two cattle dogs at her heels. She wore a pink skirt and a black floral-print blouse, and with her straw hat and fashionable sunglasses she could have been out shopping in any urban center. She greeted us warmly and invited us into her *cocina*, where she laid out a lunch of rice and beans and pots filled with mixtures of poblanos, corn, carrots, mushrooms, and broccoli—all prepared without the aid of running water. There was homemade

A lesser long-nosed bat finds a tasty meal. Like the Mexican long-nosed bat, this species relies mainly on flowering agave plants as it migrates between Mexico and the southwestern US—but it also eats nectar from flowering cacti. Its more varied diet has helped its population recover in recent years.
Barry Mansell/NPL/Minden Pictures

salsa and warm tortillas, and a two-liter bottle of blazing-red strawberry soda to wash it down.

After the meal, she took the team to see her nursery, where she was growing hundreds of agaves—tiny ones in trays, just planted a month earlier, and larger ones in neatly organized rows on the ground. These would eventually be planted out in the farm fields. She showed off her workshop, where she made a range of health- and skin-care products from local plants, which she sold to stores in Monterrey. It's one of the alternative income streams that the agave restoration effort supports, aimed at helping ensure the communities can thrive in the desert even as climate change makes traditional livelihoods like ranching less feasible. Funds from the agave project had helped build the workshop and buy packaging for the products, which Galván sold under the name Natiza—one of seven ejido-made brands marketed together as "productos naturistas del semidesierto." There was a healing balm made from neem, a shampoo that was said to pull toxins from the scalp, breath spray made with local herbs.

We drove across the state line to visit Huertecillas, another ejido, where money from the agave project had gone to purchase basic equipment to help a local women's collective make their balms and lotions more efficiently. Martina Pérez Martínez and her daughter Selena collaborated on a line of products called Arena Rosa (pink sand), which included face and body lotion and lip balm produced with essential oils, extracted using a new table-top machine, from plants they grew or collected, like rosemary and mint. A recently purchased dehydrator allowed them to make health supplements like arnica capsules. They blended other products in a just-bought KitchenAid mixer. A new workshop was under construction there, too, in order to meet sanitation standards and give the products a wider distribution. As the sun set, Martínez showed us her garden, screened-in with

mesh to help protect it from marauding wildlife—there had been a recent badger raid on an agave nursery—and watered by rows of plastic soda bottles stuck upside down in the soil.

Afterward, she served dinner—more rice and beans, plus squash stew and corn tortillas made fresh over the fire. We slept there that night. Breakfast brought another mouth-watering meal: a giant pot of huitlacoche, an edible corn fungus whose use as a meal dates back to Aztec people, and to Hopi and Zuni tribes north of the border; and mugs of a hot drink made with pinole, a powder ground from blue corn, mixed with cinnamon, cocoa, and milk.

The ejidos we visited all lay within the same general area, but between them there were subtle differences in rainfall and historic land use, enough to make for noticeable shifts in the quality of the soil, the health of the agaves, even the feel of the air. At Coyotillos, in Zacatecas, José Inocencio Moreno Mendoza took us to the field where he had planted rows and rows of agaves he and his wife had germinated from seed in the nursery. This was a different species, a non-native agave that was fast-growing and might flower in less than ten years. There were also rows of prickly pear cacti, or nopal, which are harvested as a crop, and some native plants like mesquite and a kind of small, green pumpkin that's used to make soap. But the field was dry and dusty, and many of the agaves looked like they were barely hanging on, leaves brown and shriveled. Still, Lear was confident they would survive. They were adapted to the desert, after all.

"This used to be a farm field, but when the rain stopped coming, nothing could grow," Leyequien said as she walked the rows. She hoped the agave might help keep the field from turning to desert. "If the agave can't even grow, corn can't grow. Then the soil erodes." The soil was so compacted in places that it could no longer hold water even when the rains did come. It simply

washed away. Leyequien said she was trying to help the community plant native crops to hold the soil in place, and also looking
for ways to help people access more water. She told Lear and
Ibarra she hoped to find a cheap filtration solution for the contaminated wells.

The sun was blazing down. The air, like the soil, felt desiccated.
We crossed a line of shrubby mesquite to another field, where the
soil was darker and not as dry. Here, many of the agaves looked
healthy and thriving—they were half a foot taller than the other
ones, with strong green leaves poking skyward. Some infinitesi-
mal difference from one field to the next—in rainfall, slope, perhaps past practices—seemed to have made a difference.

We visited Tanque de López, another ejido, where Bertha
Alicia Estrada Mireles and her husband, José Jesús Reyna
Martínez, had also embraced the restoration project. Martínez
showed us a burlap sack full of agave seeds they had collected.
The seeds grow in oblong pods the size of a swollen thumb; black
seeds are fertilized and can sprout into new agaves, while white
seeds aren't viable. Mireles, who wore a T-shirt adorned with
a giant butterfly, took us to the fields where they had planted
out the agaves, which were growing strong and healthy in neat
rows, interspersed with prickly pears. There was slightly more
rainfall here than at Coyotillos; the ejido also had some water
to use for irrigation. Mireles showed us her vegetable garden,
which was overflowing with cilantro, broccoli, squash, and row
after row of chiles, which she dried and sold, along with fresh
tortillas. The agave restoration project was helping with supplies for this business as well. Martínez had built a composting facility, where worms were helping turn manure into a soil
amendment. He said he hoped to sell some of it, but it was difficult and expensive to truck it to a town.

From the composting area, you could see just how little rainfall there had been; Martínez said it had barely rained here for

five years. We looked out on a reservoir designed to hold rainwater, which should have been filled all the way across to a row
of trees perhaps a hundred yards away. Instead it held what
looked like a large puddle covering maybe a fifth of the area. That
water was all the community had from natural sources until the
next rains, which typically came in the spring, a good seven
months away. The drought was causing problems for everyone.
Hungry and thirsty wildlife was coming in closer to the ejidos,
sometimes raiding their vegetable gardens. Leyequien told us
people had been sending her photos of struggling wildlife at
Mojonera, which abutted the western edge of Tanque de López;
she had seen photos of owls and coyotes choking on dirt or
rocks while desperately trying to find water. "Please, I can't see
any more of those," she said, shaking her head.

Down the dusty road, residents of Cuatro Milpas (four corn-
fields) had germinated hundreds of agaves, and these were
thriving in a nursery. More agaves on the landscape would
help the struggling wildlife, too, especially once the plants
began flowering. Agaves are keystone plants, crucial for a
whole host of species. Hummingbirds drink their nectar and
pollinate them. Woodpeckers roost in the stalks and will also
drink the nectar. Rattlesnakes make their home in the bases.
"Crows, ravens, ringtail cats, foxes, deer," Lear rattled off a list
of species that use agaves. "Tons of animals. Agaves are just
super important."

Agaves can also feed livestock, as long as there are enough
of the plants to go around, or the animals are only eating some
of the leaves. Harvesting the outer leaves of the agave does not
harm the plant; in fact, it can stimulate the plant to produce
"pups," clonal babies that emerge from tendrils and will grow
into new agaves. This is one of two ways that agaves reproduce, but because it produces a genetically identical plant, it
doesn't preserve the genetic diversity necessary for the species'

survival. For that, it needs to reproduce via pollinated seeds, which is where the bats and other pollinators come in.

Our host in Cuatro Milpas, Aurelio Gaytán Mendoza, wanted to show the team a small wheeled mill, bright red and roughly the size of a motorcycle, purchased with funds from the restoration project. It was enabling him to feed Russian thistle—the invasive plant otherwise known as tumbleweed—to cattle. Cows won't eat the plant in the field but can digest it when it is chopped into tiny pieces. Using the milled tumbleweed as extra food for livestock helped to remove invasive plants from the landscape and left the agaves out to bloom. Mendoza and his sons had filled the back of a pickup truck with tumbleweed, which they called *maroma*, Spanish for "rope." They unloaded it beneath a shaded structure built from mud-brick walls and a sheet-metal roof. They spread out a tarp, started up the machine, and fed the maroma into the mill, until the whole truckload was ground up on the tarp.

Watching the ejidatarios grind weeds into cattle fodder, it was easy to forget all about the bats. We were hundreds of miles from the little winged mammals that Lear and Ibarra had devoted their careers to saving, which would at that moment have been sleeping the day away in a warm cave near the capital. But the bat biologists grinned as tumbleweed particles filled the air. Their project was materializing before their eyes. "As biologists, we learn about so many things," Ibarra said. "I'm learning about livestock food!"

Lear nodded. "Who knew bat biologists would be learning to feed cattle?"

Biologist Kristen Lear of Bat Conservation International plants an agave in northeastern Mexico. The organization's Agave Restoration Initiative aims to restore a crucial food source for endangered bats. *Chris Gallaway/ Bat Conservation International*

The men stuffed some of the ground maroma into a sack and turned to a wheelbarrow containing a stack of large agave leaves, several feet long apiece, sustainably harvested from the outer edges of the plants. This was another way the agave could benefit the ejidatarios as well as the bats. Mendoza fed the leaves into the grinder, and a refreshing smell filled the air, which was still thick with tumbleweed. It was the olfactory equivalent of a glass of ice water on a sizzling summer day. Once they had produced a pile of watery agave pulp, one of Mendoza's sons scooped some into a sack and mixed it with molasses. They took a sack of each type of milled feed, the maroma and the agave, to a fenced-in cattle enclosure, where the animals eagerly devoured their meal—no agaves or other native plants destroyed in the process.

— – – —

Driving back to Monterrey on the main highway after our ejido tour, stomachs stretched from eating a meal at every ejido we visited, we passed a section of road where the median was planted with succulents, including some large agaves that had just finished flowering. I asked Ibarra and Lear if they felt hopeful every time they saw an agave, anywhere. "Yes!" Lear said. "I really do. And these ones flowered, so they probably had bats feeding on them." She paused for a moment, thinking about the bats winging in to feed along this perilous highway median. "I just hope they didn't get hit by a truck."

The drive, which had taken four hours on a quiet Sunday, took seven on a Wednesday. We sat in a miles-long line of semis, many of them headed for the border. As we crossed the mountains west of the city, snaking beneath jagged slabs of limestone, it started to pour. Fat raindrops pounded the Toyota's roof. "Do you think it's raining at the ejidos?" I wondered.

"Probably not," Ibarra said.

I parted ways with the two bat biologists in Monterrey. On the way home, my flight to Houston was briefly delayed

when a baggage handler noticed that an external panel on the plane was loose. The pilot, perhaps oversharing, got on the PA and explained that mechanics had re-attached the panel and were just waiting for someone to seal it up with tape. I burst out laughing. They were taping the airplane back together? I thought it couldn't possibly be true, but a few minutes later I saw the mechanic beneath my window with a roll of what looked like metallic gaffer's tape. (It turns out this is, indeed, how planes are sometimes repaired. Solutions are often much simpler and more low-tech than we imagine.) I turned to the man seated next to me and we shared a moment of wonder tinged with concern. It seemed somehow fitting: Suspended between awe and anxiety was how I spent much of my time reporting on conservation. This trip to see the agave restoration project was no different.

I thought about how many interconnected people and systems and objects my journey home depended on: mechanics, pilots, fuel supply chains, security screenings, binational governmental agreements, radar, tape. The bats' safe passage from south to north and back, once dependent entirely on natural cycles, now similarly relied on the precise calibration of seemingly disparate human creations: mines and margaritas; a changed climate; dehydrators, blenders, tumbleweed grinders; borders, sanctuaries, restoration projects; partnerships between scientists and desert farmers. As we taxied down the runway, I hoped that the tape would hold—and the agaves would survive to bloom.

A Wilder Europe: Bears, Rewilding, and the Beauty of Coexistence

On a late-summer morning in Italy's Central Apennines, the sun was already blazing down. I was sweating and feeling about a hundred years old as I scrambled to keep up with the fit young team of European students who were looking at their phones and debating the best path to our destination. "I think we're going to have to go down and then up the other side," said a tall Swedish guy named David. I looked where he was pointing. We were standing at the edge of a shallow gash in the hillside, where the land fell away at a near-vertical slope on soft, slippery dirt dotted with thorny dog rose bushes. David was navigating toward a set of GPS coordinates, but the trail seemed to have vanished. I had been blindly following the group, assuming

---

Previous Spread: A mama Marsican bear and her cub in Italy's Central Apennines. Just an hour and a half from Rome, this mountain chain has been called "the wild heart of Italy." The mountains here are home to mammals including wolves, deer, boar, and these critically endangered bears, which live nowhere else in the world. *Bruno D'Amicis*

they knew the route rather than simply the end point. Now, apparently, there was nowhere to go but down.

I tried not to lose my sense of humor as I followed a French field biologist named Sarah down the incline, bracing my feet against tree trunks and cursing the thorns that ripped at my clothes. At least down in the ravine it was shady. The scramble up the far side was easier, and from there, finally, we could see the trail we were supposed to have been on all along. To the east, the village of Ortona dei Marsi sat perched beneath the limestone formations of the Central Apennines, rounded and low in the foreground before giving way to more jagged peaks farther afield. The Central Apennines are less dramatic, less snowy, and less internationally famed than the Alps. (The highest peak in the chain, the Gran Sasso, rises to 9,553 feet, just under 3,000 meters, but most of the mountains stand below 6,000 feet.) They are also arguably less tame. Under two hours' drive from Rome, this is one of the wildest parts of Italy, home to a significant portion of the country's gray wolf population and a critically endangered bear that lives nowhere else.

When we think about large mammals moving across a landscape, we tend to picture sweeping vistas: zebra grazing the African savanna, or Amur tigers prowling the Russian steppes, or caribou crossing the tundra. We don't think about the industrialized environs of Western Europe. At least I didn't. But that is changing. A "rewilding" movement has been taking shape around the continent—an effort to restore natural processes and recover wildlife populations so that ecosystems can adapt and thrive. Rewilding is largely about "creating the enabling conditions," as the organization Rewilding Europe puts it, so that nature can regenerate.

The influential American conservation biologist (and founder of the radical environmental group Earth First!) Dave Foreman, who died in 2022, coined the term "rewilding" in the

early 1990s. He and others, including biologists Michael Soulé and Reed Noss, built a restoration vision around what they called "the three C's," cores, carnivores, and corridors—core protected areas, linked by corridors, with keystone predators such as wolves and grizzlies helping regulate the whole ecosystem. These ideas have been the basis for successful conservation programs in the US, including the reintroduction of wolves to Yellowstone National Park—where their absence thanks to human obliteration had altered the landscape and their return allowed trees to regrow along streams and rivers, which enabled beavers to come back, which created new wetland habitat for a range of species, and so on.

The rewilding idea took shape in Europe around the same time in the late twentieth century, with a Dutch experiment on a previously drained delta of the Rhine River, known as the Oostvaardersplassen, or OVP, that "involved the reassembly of a guild of large herbivores, including 'wilded' horses and cattle and red deer, to create a Serengeti-like landscape," as one account explains it. (The OVP project was and remains extremely controversial, as the animals, whose herds have grown in size, are left to fend for themselves—and many starve during bitter winters.) More recently, though, as land-use change and climate change collide, rewilding has gained momentum across multiple continents to restore lost wildness, make natural landscapes more resilient, and foster coexistence between humans and other species. Rewilding Europe promotes efforts across thirteen countries, covering landscapes

---

Marsican bears, a subspecies of brown bear, live only in the Central Apennines. They once occupied a much larger swath of Italy but deforestation and hunting decimated their population. Only about sixty individuals remain today—though their birth rates are rising. Providing safe passage for the bears between protected areas will be crucial to their continued survival. *Bruno D'Amicis*

from Scottish moors to a Polish delta to Swedish Lapland, and involving recovery efforts for king vultures, European bison, Balkan chamois, white-tailed eagles, and dozens of other species. I had come to Italy not, as many people do, on a pilgrimage in search of art or fashion or pizza. I had come to learn about the bear.

The Marsican brown bear (*Ursus arctos marsicanus*) is a subspecies of the Eurasian brown bear, which itself is a subspecies of brown bear. (Grizzly bears in North America are also considered to be a subspecies of brown bear.) Marsican bears once occupied a much larger swath of Italy, from the Sibylline Mountains at the north end of the Apennines, near Perugia, down to Calabria in the country's south. But over decades and centuries, deforestation for agriculture and the access for hunters that this land conversion wrought—not to mention the advent of guns—spelled disaster for the animals. The last Marsican bear in the Sibyllines is believed to have been shot in 1870. Over time the bears' habitat shrunk further and further, until they became isolated in Parco Nazionale d'Abruzzo, Lazio e Molise, a national park established a century ago.

All of the subspecies' reproductive females currently reside in that park, while the males range across a series of adjacent and nearby protected forests. The entire population of Marsican brown bears numbers only around sixty. During the last decade, though, their birth rates have been coming up, and an increasing number of bears have been moving between these parks and reserves, using trails and routes between the protected areas.

Local conservation groups have sprung up to help the bears, working with park officials and more recently the EU to create "coexistence corridors"—safe passageways to help animals move between the forests and recolonize the surrounding natural areas. There are regional parks to the north and west, a series of reserves to the southeast, and another national park

to the east, but we aren't exactly talking about a Denali, or even a Yosemite. "In Italy, you can be on top of a mountain and you'll always see a town," Fabrizio Cordischi, an Abruzzo native, told me. "It's so small a country, so coexistence is so important." We were standing on a hillside looking across a valley at a herd of red deer, one of the largest species of deer in the world, on the opposite slope. It was the rutting season, and the males were bugling, making their strange trumpet calls to attract mates. Once our eyes had adjusted to spotting the animals against their rocky backdrop, we could see some males batting antlers. The females seemed oblivious to the males' questionable charm.

Cordischi was the field manager for the local rewilding project, called Rewilding Apennines, which was confronting a similar problem as in Costa Rica—how to build safe passage for animals across human landscapes—but in a very different context, in a wealthy and intensely urbanized country. Italy's gross national income per capita is three times that of Costa Rica's, and its population density is nearly double. As of 2023, there were about 506 people per square mile in Italy, compared to just 267 in Costa Rica. And though Italy is about six times bigger in size than Costa Rica, it contains thirty-seven times more paved roads. While most tourists who visit Costa Rica are drawn there in large part by the wildlife, Italy is a different beast—a haven for fashion, art, coffee, and other urban preoccupations. Sure, there are stunning backdrops and lots of opportunities for outdoor recreation—the Dolomites, for one, are a popular hiking destination—but it's not a country that most travelers generally associate with wide-open spaces and wildlife. Most of Italy was long ago built for humans.

Still, the entire country is wilder than it might seem. As of a 2021 survey, about 2,400 wolves reside in Italy. In the roughly 2,000-square-mile section of the Central Apennines where

Rewilding Apennines works, there are about 400 wolves—or a sixth of the country's wolf population living in about a sixtieth of its land area. There are also red deer, wild boar, Apennine chamois, European wildcats, and griffon vultures, among other iconic species. Over the centuries, though, Italians cleared massive swaths of forest in these mountains for agriculture, using the timber for heating, construction, and railways. In the 1950s and 1960s, the government began widespread reforestation efforts, planting seedlings across millions of hectares of land. You can still find remnant old-growth beech and maple forests, though many of the trees today are conifers brought in by the Forestry Corps to help prevent soil erosion and employ rural residents during the difficult post-war era.

The forests eventually regrew, but the villages began emptying out. Industrialization and urbanization sent younger people fleeing the mountains for jobs in the cities. As one recent Italian research paper put it, "Since the 1950s, the Apennines were particularly exposed to a demographic decrease that was fast, virulent and widespread." This phenomenon of depopulation has led to a strange mix of circumstances for the local wildlife. While more sparsely populated villages give large mammals more space to roam, there's a downside: An aging population without younger residents has let outdated cultural attitudes about wildlife—that they are a threat, rather than an asset—persist.

In a paper published in 2021, a group of scientists coined the term "anthropogenic resistance." It refers to the ways in which "human behaviors influence connectivity." Conservation efforts, they wrote, often considered "resistance surfaces" across a landscape—the kinds of things Chris Beirne, in Costa Rica, was adding to his maps of possible wildlife pathways. But they should also consider, the scientists argued, "the direct effects of human behaviors on species' movement." While there might

be a physical pathway across an area, that did not mean wildlife could necessarily use it; human activities in that area—hunting or trapping, or drug smuggling, or war—might make it a dangerous or inaccessible corridor. Arash Ghoddousi, an Iranian biologist now at Wageningen University in the Netherlands and a coauthor of the study, had been working on a project to understand connectivity for leopards between two national parks in Iran. Inspired by work about jaguar connectivity in the Americas, which used surveys to understand people's attitudes toward the carnivores, he began collecting similar data. "Let's see what our habitat models are showing us about where are habitat patches, where could be corridors, but let's also incorporate conflict risk," Ghoddousi said. He combined a map of risk from human attitudes with a map of potential corridors and realized that the corridors went through "certain areas of conflict that could be counterintuitive, leaving leopards to areas where they might be killed."

Today, some threats to the Marsican brown bears are structural: Between 1970 and 2022, thirteen bears were hit by cars, largely on the narrow, twisting roads that connect the Apennines' villages. Another seven were hit by trains. But some threats are cultural, caused by anthropogenic resistance. In 2014, near Pettorano, an elderly man shot and killed a bear that was approaching his chicken coop. The bears are protected; it's illegal to kill them. The killing triggered outrage, and the man ultimately paid a hefty fine—though the case took seven years to wind through the courts. (In the summer of 2024, another bear was shot and killed—a beloved sow who was raising two cubs and whom locals had nicknamed Amarena after a variety

Next Spread: A herd of Apennine chamois, a species of goat-antelope endemic to the region, moves across the mountains of Majella National Park in Abruzzo, Italy. Wolves and many other iconic species live in this area. *Bruno D'Amicis*

of black cherry she especially liked. This time, a community uproar ensued, and even the local governor and the country's environment minister weighed in to express their dismay.)

Mario Cipollone still gets visibly agitated when he talks about the 2014 bear killing—as he did one evening over foraged-mushroom pasta in the picturesque village of Pettorano sul Gizio, which sits inside the boundaries of a regional forest preserve. Cipollone and his wife, Angela Tavone, have lived in Pettorano since 2019, and when I visited in 2022 they were temporarily crammed into a tiny apartment with their young son Lorenzo while they remodeled a larger house higher up the winding village path. Together, they run Rewilding Apennines, whose mission is to protect and grow the wildlife populations there and create sustainable models for thriving local economies that capitalize on—rather than force out—nature. Their aim is to cultivate two types of connectivity—one physical, the other socioeconomic.

Pettorano sul Gizio, built into a steep hillside, was a thriving town of roughly 5,000 people in the early twentieth century. Today, around 1,300 people live in Pettorano, but three-quarters of them live outside the main village, in what locals call the "suburbs" but Americans would more likely call a rural area, along the road to Sulmona (population 25,000). Because most people shop in Sulmona, the businesses that used to dot Pettorano have largely closed; today, there are two restaurants, two bars, a pharmacy, and a small convenience store at the base of the vertical village. When I visited, in mid-September, the shop and one restaurant were shuttered for vacation, and many of the homes were empty. Tavone told me that a lot of the homes belong to people whose parents and grandparents once lived there full-time; today, the descendants are far more likely to live in cities like Rome and come to Pettorano on summer holidays.

In choosing to build a life in Pettorano, Tavone and Cipollone made a conscious commitment to rebooting the local economy as part of the rewilding mission. For the 2022 holidays, Rewilding Apennines and partners from the various parks put together a gift box, the "Bear-Smart Box," full of locally produced items. Available for delivery in Europe through the website Broozy, the box's goodies, according to the site, traced "the gradual expansion of the Marsican brown bear which, from the ancient forests and the most inaccessible places of the Abruzzo, Lazio and Molise National Park, as the population grew, began to colonize other territories outside the core area where he had always lived, crossing the ecological corridors, large territories of connection between one protected area and another." The products included artisan sheep cheese, small-batch cider, black cherry liqueur, chocolate-hazelnut nougat, and a pesto made from mugnoli, a local vegetable related to broccoli.

It was just one tiny initiative on a steadily growing list of tiny initiatives that were collectively amounting to something significant. An ecotourism company called Wildlife Adventures received loans from a Rewilding Europe fund aimed at helping small, nature-friendly businesses. The student volunteer program helps generate income for local bars and restaurants by bringing forty students a year for three-month stints. And Rewilding Apennines built a vulture-feeding station where farmers can bring dead livestock—which they would otherwise pay to incinerate—that helps farmers save money while also promoting conservation. The group's "enterprise officer," Valerio Reale, had also recently moved to Pettorano, where he was working in part with local "bear-smart" artisanal businesses. Reale and his wife, who previously worked on an organic farm in the UK, hoped to start up a bakery in Pettorano. (Once upon a time, the village housed five bakeries, all with "common ovens" that villagers could use to bake their own bread.)

In its efforts to create coexistence corridors for wildlife and people, no idea is too small or too out there for Rewilding Apennines. Coexistence means, on some level, treating wildlife as your neighbors. You might not bring them a pie or give them a stick of butter, but you should consider how your own actions help or hinder their ability to muddle through.

Along Abruzzo's winding roads, the group's staff and volunteers have installed devices that make loud noises like fireworks when cars are near—designed to scare bears and avert deadly collisions. They have also installed electric fencing around local beekeepers' hives, to keep the bears out and the anti-bear sentiment that might result from honey theft at bay. They have taken down barbed-wire fencing that was installed in the 1950s and '60s to protect the newly planted trees. "Many of those plantations didn't grow," said Tavone. "Results weren't the same everywhere. But a common theme was that no one removed the barbed wire that was used to protect the seedlings." The team has been removing the wire since 2018 and has taken out more than 110 kilometers (about 70 miles) of fencing. "No one will go there and rebuild these fences," she said. "It's like a permanent restoration action."

Rewilding Apennines workers have also installed some new fences: They've put electric fencing around apple orchards to reduce the risk of conflict from bears helping themselves to farmers' fruit. But they haven't left the bears high and dry. In a move that seems bizarre from an American perspective, they have pruned some fruit trees—apples, wild cherry, and wild pear—specifically to feed bears. Once upon a time, said Tavone, "agriculture reached high up in the mountains. But after the 1950s and '60s it was abandoned, as people moved to the cities. A large number of these fruit trees are still alive and are productive." Keeping those trees healthy and yielding food that bears and other wildlife can access may prevent the animals

from raiding still-in-use orchards elsewhere. "In order to not lose this important food source for wildlife, we are going to the mountains to check the state of health of these plants, to free them of competition from other plants, and prune them in autumn, to guarantee additional food for wildlife, food that is far away from the towns," Tavone told me.

Rewilding Apennines has a small staff. But the group runs largely through the manpower of an army of volunteers, mostly students from around Europe and the UK, like the ones I was hiking with, who are taking a gap year or need an internship to graduate. These unpaid interns patrol the trails, looking for signs of bears—and also of other species of interest for regulating the ecosystem, like European wildcats and griffon vultures. The interns monitor roads for wildlife killed by cars, as a way to identify problem points and work to find solutions. They help maintain camera traps—a network of wildlife cameras installed throughout the region, mostly strapped to trees in the forest. They walk transects beneath wind turbines, looking for any flying creatures that might have been killed by the blades. They participate in vulture counts run by the national parks. And they do a host of other activities, like looking for uncovered wells out on the hillsides—which is what we were doing the morning we got lost.

Built throughout the nineteenth and twentieth centuries, the wells were used to supply water for cattle grazing. Older ones contained "clinging points" or even staircases, but those constructed in the 1960s and '70s, said Cipollone, "are bigger, deeper concrete reservoirs without any escape ways." Many of these wells were not covered; they were simply holes in the ground, in remote areas, "that could really turn into traps for wildlife." In recent years, five bears have died in wells—two mothers and their cubs. So, Rewilding Apennines set out to find and cover every single opening. Each time someone finds

one, they mark its GPS coordinates and measure it. Then they recruit a local metal fabricator to custom-cut a grate, and they carry the grate to the wall and install it. So far, they have covered twenty-seven open shafts.

On the Monday morning when I tagged along, the volunteers had set out from the tiny village of Ortona dei Marsi to take measurements of one open well that they'd learned about, and to scout for any others. Eventually, we arrived at the well, though what should have been perhaps a twenty-minute walk had taken more than three times that long. After measuring the concrete opening, the group decided to split up to save time. My small faction followed a path to the west, back toward Ortona; we found two more open wells along the way. Grace, a British student who was studying environmental photography, took measurements while David, the Swede, marked the coordinates. Eventually we reached the road, and just a few moments later a white pickup truck bearing the organization's logo magically appeared and pulled over to collect us. There were already four men inside, but we squeezed in and headed up a narrow dirt path. They had just come from picking up a grate that a local welder had made, and the next task was to install it.

It was midday by this point and the sun was searing. Everyone in the group grabbed something that needed carrying— toolboxes, a drill—and two of the guys hoisted the metal grate up onto their shoulders. We trudged uphill across a shadeless meadow until we found the well, and the group set to work drilling holes in the concrete and securing the grate on top. I drank the last of my water in a meager shade patch behind a dog rose, listening to the whine of metal cutting concrete. I had to admit that all this seemed like a lot of effort just to cover up a manhole on a hillside. Then again, when you're working to save an animal with just five dozen individuals left on Earth, you become keenly aware of how much each life counts, and how

any intervention, however farfetched or trivial it seems, might help fend off extinction.

When the work was finally done, we piled back into the truck, with Cordischi, the field officer, at the wheel. We dropped part of the team back in Ortona and headed to a nearby village for pizza. (Even if you're not in Italy expressly for the food, you still have to eat!)

— - - —

One day, I sat in on a presentation about how to recognize various "signs of presence" of bear, including overturned rocks— in early summer, bears flip them over to hunt for ants. About fifteen twentysomethings crammed into the living room of a house that Rewilding Apennines uses to lodge its volunteers. They perched on chairs, sprawled across a futon, or sat cross-legged on the floor, while a German ecology graduate student who works as a GIS analyst for the group talked in detail about poop. "You'll see lots of pictures, mostly of bear feces," Jan-Niklas Trei said to kick off his talk. He showed photos of bear poop at different times of year, varying in appearance depending on what the animals were mostly eating that season: fruit, acorns, insects. He explained that if they saw bear feces while on a trail, the volunteers were to mark it with a ruler, take a picture, and upload it to EarthRanger, the app they use to record field observations.

As the volunteers took in the material and padded to the kitchen for coffee refills, I thought about how all these budding naturalists could have been working to get rich at tech companies, but instead they were devoting their time to helping save bears and restore an ecosystem. It also struck me that these bears just trying to make their way in the world were oblivious to the dedication all these people were bringing to help them. They would never know just how many people were lovingly studying their poop.

In the afternoon, I set off with Jan and Sarah, the French biologist, and Jan's sister, Mina, who was visiting from Germany. The scientists were going to collect memory cards and swap out batteries on some of the camera traps. We zoomed along the winding, sometimes harrowingly narrow roads in a pickup that felt menacingly large for the surroundings. The camera locations had been arranged based on a grid system, and some of them were up trackless slopes so steep that Mina and I stayed below so as not to slow down Jan and Sarah, who'd been doing this all summer and could by now scramble up the mountainsides like goats. At one location, there was a skinny dirt road that wound up the hill in a series of hairpins, with a near-vertical drop down which you'd careen if the driver miscalculated by just a couple of feet.

We drove until the road became impassable, and then stopped to take in the view. More than a thousand feet below us in the valley, the Apennines attraction of Lago di Scanno sparkled in the afternoon light. A few late-summer tourists were out in boats, which just looked like dots from up on the mountain. We set off on foot down the slope, following a wildlife trail through the leaf litter. I found lots of "signs of presence": scat from deer, fox, even horse, part of a heartbreakingly exploited EU policy that pays farmers for livestock and has given rise to an immoral practice of farmers buying horses and then setting them loose on steep mountainsides like this one, where they are left to die. I looked for signs of bear but didn't find any.

The camera trap was attached to a beech tree a few feet off the ground. Sarah opened it and began scrolling through the photos, just to be sure it was working. She let out a yelp of

An Apennine wolf caught on camera, on a hillside high above Lago di Scanno. Rewilding Apennines maintains a series of wildlife cameras throughout the area to monitor wolves, bears, and other animals. *Rewilding Apennines*

excitement and we bent down to see what she was looking at. In a series of photos taken just three days earlier, there was a wolf. We stood there for a few moments, digesting the discovery. It wasn't exactly surprising; I knew there were wolves living here. But seeing its photo, snapped so recently in the precise spot where my feet were right then planted, made the whole thing real. Large, wild mammals were roaming this landscape, and by working to make the matrix more welcoming and change attitudes to reduce anthropogenic resistance, the rewilding team intended to keep it that way.

Mario Cipollone and a few others were headed to Canada the following week, to learn from bear-smart communities in British Columbia and see what might translate for Italy. Back at home, Tavone and others were planning outreach events for the coming year—to help connect locals with nature, talk about things like human-wolf coexistence, and involve communities in the rewilding efforts. There was no time to lose. In the spring of 2023, park rangers and Rewilding Apennines staff out doing some monitoring in an important corridor between two protected areas discovered the poisoned carcasses of nine wolves, five griffon vultures, and two ravens. An entire wolf pack had been lost. Still, the rewilding efforts continued. They were gearing up to reintroduce griffon vultures and Apennine chamois to the ecosystem. They were working on projects large and small; it all mattered.

— – – —

I left Pettorano early one morning and boarded a train in Sulmona. I was headed to the region of Trentino–Alto Adige, in the Alps. I passed through Bologna and Verona, cities bustling with people—the vast majority of whom never gave a moment's thought to the existence of bears or wolves or wildcats tantalizingly near. It was almost dark when I reached my final destination, a station called Mezzocorona, just outside the town of

San Michele all'Adige. Andrea Corradini, a PhD student working on large carnivore ecology, met me at the station and we went to a nearby bar for cocktails—delicious concoctions made with a local elderberry liquor—while we waited for his advisor, Francesca Cagnacci, to arrive.

Corradini and Cagnacci had recently collaborated on a study, with a handful of other researchers, that used data from the outdoor activity tracking app Strava to show that brown bears in the region avoided places where there was a high density of human activity, even if the area represented good habitat from an ecological standpoint. Here in the country's north, surrounded by the signature jagged Alpine peaks, the valleys teem with people and intensive agriculture. Italy is the world's sixth-largest producer of apples, and the second-largest exporter, after China—and Trentino's valleys are the epicenter. Everywhere you look are fenced-in orchards, jam-packed with trees growing apples—mostly Golden Delicious, often bound for markets in Asia. The valleys are also crisscrossed by a dense network of roads and train tracks. It's not exactly the kind of place you'd expect to run into a bear. But they live here.

Unlike the Marsican bears in the Apennines, the bears that live in Trentino are not technically natives—in the sense that they're not descended from the region's original ursine occupants. Brown bears once roamed much of Europe, and they lived across many parts of Italy into the eighteenth century. But at the end of the 1700s, large-scale deforestation for agriculture shrank their habitat and also gave hunters access to what had once been remote wilderness. It wasn't long before Italian bears were pretty much toast in much of the country. In the eastern Alps, scientists estimated that for nearly 150 years, from 1850 to 1995, bear numbers declined by 10 percent a decade, thanks to both legal and illegal hunting—including by farmers to protect livestock, beekeepers to protect hives,

and hunters in pursuit of bounties paid by local governments. By 1997 there were just three bears remaining in the Italian Alps; by 2000 there was only one, an elderly and partially blind male.

The Large Carnivore Initiative for Europe—a group of scientists within the IUCN (the International Union for Conservation of Nature, a giant network of scientists, governments, and NGOs from around the world)—recommended the reintroduction of bears into local areas where they could thrive and ultimately, hopefully, merge across international borders into large, healthy metapopulations—those connected groups that can mix and mingle to breed. Ten bears were caught in Slovenia and relocated to Adamello-Brenta Nature Park, in western Trentino, between 1999 and 2002. The aim was to build a metapopulation that ranged from Italy east to Slovenia and throughout the Dinaric Alps all the way to Albania. The relocated bears thrived in the park and spread outward, colonizing other forested areas of western Trentino. But they never made it any farther east. There were likely somewhere between eighty and ninety bears in the area as of 2020, and the population was listed as critically endangered.

"The vast infrastructural system of the Adige basin effectively constitutes a connectivity barrier for many animal species," Corradini and his colleagues wrote in 2021. Still, they wanted to see if the presence of people—the humans themselves, not simply human infrastructure like roads—made a difference to how bears used the landscape. As they put it, "Humans profoundly affect animal distributions by directly competing for

An apple orchard in Trentino, Italy. The country is the world's sixth-largest producer of apples, and only China exports more of the fruit. Trentino's valleys are filled with fenced-in orchards, many growing Golden Delicious apples for export to Asia. *Luciano Gaudenzio/Photo FVG*

space, not only transforming, but actively using their habitat."
It was a kind of unintentional anthropogenic resistance.

No one had used Strava data to study ecology before, but it
was a perfect fit. Strava is a tool where users track their outdoor activity and share it on a social media platform, nam-
ing their bike rides or hikes and commenting on each other's
pursuits. "Group therapy," "fried ham(strings) for breakfast,"
"Does it hurt when I do THIS?" and "I fell asleep but just kept
pedaling" are some quintessential Strava titles (okay, full dis-
closure, they're all rides posted by my husband). Users com-
ment on each other's efforts, leaving notes: "Yaasss!" or "Wish
I were there with you." The tracker creates a linear map of
your route, so some users even create Strava art—running or
biking a route that resembles the *Mona Lisa*, say, or the shape
of a bear, or, perhaps unsurprisingly, penises. Lots of penises.
Strava combines its users' data—billions of records of outdoor
recreation—into something called the Global Heatmap, on
which heavily used trails and areas light up, much like cities do
on nighttime images of Earth from space.

Corradini and Cagnacci used the Global Heatmap to look at
what they call "functional human disturbance," the presence
of humans as they move through and use a landscape. They
used the Strava data to create something called a Cumulative
Outdoor Activity Index, or COI, a measure of "active competition for space between humans and wildlife." The researchers took this information and compared it to GPS data from
twelve Alpine brown bears that had been radio-collared
between 2011 and 2019. They also set out camera traps to
document the presence of both people and wildlife along
trails and forest roads.

There are worse places to do ecological research than
Trentino–Alto Adige, in the Italian Alps. *Andrea Corradini*

The study found that within the bears' home ranges, the animals, in scientist-speak, "selected for steep areas and high canopy cover and strongly avoided areas with high density of functional disturbance." In other words, the bears steered clear of trails where lots of humans were out hiking and biking and high-fiving each other about it later on Strava.

"In light of these findings, the establishment of a long-term, viable Alpine-Dinaric brown bear metapopulation may be difficult to achieve," the paper concluded. The bears' long-term survival was at risk as the animals "continue to search for space in this increasingly complex and expanding matrix" of human use. In addition to a boatload of busy highways and high-traffic train lines, the sheer mass of people out enjoying nature, some (like me) maybe even hoping to catch sight of a brown bear, was limiting the bears' ability to move—and jeopardizing their prospect of expanding eastward to connect with their Slovenian cousins.

The paper's findings presented a familiar but increasingly urgent quandary for conservation. Getting more people outdoors and into nature has long been seen as key to convincing them to value and protect it—but not if the cost is loving natural areas to death. The pandemic made this problem viscerally clear in many places, including where I live in Colorado, when outdoor recreation was the only real way we could get together with friends and neighbors. The trailhead at the end of my street, previously used mainly by locals and where for a decade and a half I rarely saw a car parked, suddenly became so popular that cars lined the neighborhood and single-track trails eroded into wide paths. The state park down the road had to institute a reserved-parking system, and other popular hiking areas began offering both timed parking and shuttle service from town.

It was the same, more or less, in the Alps, thanks to a growing trend of nature-based recreation over several decades.

"Whereas in the '80s, '90s, the main kind of outdoor activity was linked to mass tourism—ski resorts—or maybe people would go for a stroll, now it's way more pervasive use of the mountains and outdoors that we observe," Francesca Cagnacci had told me from her car during the pandemic, shortly after Italy had emerged from Covid lockdown. "This is actually great, but the downside is that this can easily become a way to use the outdoors that is similar to an entertainment park. People may not be aware that the woods are inhabited by wildlife—vibrating life behind the scenes that has to be appreciated."

Italy's pandemic lockdown offered a rare chance to further study how the presence of humans alters bear behavior. From March 9, 2020, through May 18, 2020, the country was under the most intense lockdown in all of Europe. Residents had to stay within 200 meters of their homes, and no outdoor recreation was allowed. The sudden absence of humans from the landscape presented a unique opportunity to look at bear behavior. Corradini looked at cases of bear-related damages reported to the authorities between those same dates, March 9 and May 18, but over a five-year period ending with the pandemic.

He found that there were far more reports of damage to human properties—categories including "poultry, garbage, building, and beehive"—in 2020 than in most of the other years. The data showed that bears "entered human-dominated spaces" more frequently during the lockdown. As he and others wrote, "With no sensory stimulation associated with human activity, bears emerged from hibernation to find undisturbed spaces and availability of otherwise little accessible resources, prompting them to take advantage of these unexpected opportunities." I imagined the bears coming out of their dens and finding, as opposed to every other spring, virtually no signs of humans. And what must they have thought in mid-May when, finally, the roads were again full of cars, and people were everywhere?

Without their main competitor—us—during that two-month stretch, bears spread across the landscape in a more even way. Corradini told me that one radio-collared male bear even crossed—finally!—the major highway bisecting Trentino. From there, he could have lit out for Slovenia. But for whatever reason, he simply crossed back again and stayed closer to home.

— – – —

Corradini and I sipped the last drops of our cocktails and went to find Cagnacci, who had just finished up some work meetings even though it was after 9:00 p.m. The three of us drove north in the dark through the Adige Valley to the town of Livo, home to Cagnacci's "field station." I'd been picturing a rustic structure on the edge of a forest, so I was surprised when we pulled up to a very nice-looking house in a village, which she had rented for the summer field season. Up a flight of marble stairs, a large kitchen bustled with activity, as the field assistants put the finishing touches on a feast they had prepared for our arrival. There was local cheese, a lentil stew and a cabbage dish made with local produce, a type of polenta made with local rye. There was locally brewed beer, limoncello that someone's mother had made, and homemade chocolate cake for dessert. This was my kind of fieldwork.

Cagnacci's work is focused more on basic science than on conservation, but it has important conservation implications. She's a pioneer in the field of movement ecology, which has the potential to help scientists make better conservation decisions, by showing exactly where animals go, how they move across ecosystems, and what kinds of barriers stand in their way.

Cagnacci was working in an Alpine area in South Tyrol, the northern region of Italy abutting Austria, where residents mainly speak German and it feels like you have already crossed the border to a different country. As we drove north from Livo the next morning, we traversed some imaginary line

and suddenly the architecture was distinctly Austrian, the road signs in both Italian and German. We stopped in the town of Lauregno/Laurien and parked on a rise overlooking a church and an elementary school. Cagnacci wanted to show me this spot because it influenced how she thought about connectivity. From the vantage point, you could see several different valleys and mountain ranges. "I never saw, in all this traveling around Trentino and Alto Adige these last years, a place that I would really define as a crossroad of connectivity like this one," she said, as we looked out on a panorama of mountains and valleys spreading into the distance. "This is an in-between place. From this single spot, you see so many different systems."

She was thinking about the various landscape features in the places that surrounded us—higher peaks and gentler hills, rocky slopes and green meadows, dense forests and tundra—and also the ways in which humans used these places. There were apple orchards, open pastures, woods that were selectively logged. "When I think of ecological connectivity and I imagine a hub, like a hotspot, this will be that place." Like Andy Gonzalez, the Montreal ecologist, Cagnacci was looking at the landscape from the point of view of an animal. "Where are the corridors?" she wondered aloud. "Where are the resources?" She was trying to "read" the landscape the way you might read a book or a scientific study, piecing together its important details and themes, to see it through the eyes of a bear, or a deer, or a wolf.

Cagnacci had initially become interested in studying the movement of roe deer, a much smaller species than the red deer I'd seen in the Apennines, to see how their activity was related to transmission of ticks. Scientists had thought roe deer were a static species, one that moves around locally but doesn't migrate by season. But one study had hinted at possible seasonal migrations. Cagnacci put radio collars on deer in the southern part of Trentino, and her research showed that roe

deer are what's known as a partial migratory species, which
means some animals migrate and others stay put. Out tracking
deer in Trentino's forests, Cagnacci also noticed that "things
were changing super fast, for the better and worse." Improved
land management practices in the region were helping wild species thrive, but climatic changes were also altering the assortment of mammals that shared the landscape. Meanwhile, signs
of warming were becoming more and more evident. The snow
line was moving higher up the mountainside. Snowfall was
becoming less regular.

In a subsequent study, Cagnacci and colleagues put radio collars on 104 animals from three species of European deer to get
a sense of how migration patterns differed between individual
deer. They found that among populations that were partially
migratory, the patterns were inconsistent; there were "inter-
mediate space-use strategies" that fell somewhere between
complete seasonal migration and constant residence in one
place. They called such neither-here-nor-there landscape uses
"a residency-to-migration continuum," and they predicted that
both climate change and land-use change would cause populations to shift along the continuum, migrating more or staying
put more depending on specific conditions.

As species move around differently due to climate change,
those variations will affect whole ecosystems. Cagnacci wondered not only whether certain animals might or might not be
present in a landscape, but how the species that were there
would relate to each other in terms of food webs—how the complexity of those webs would change and what that would mean
down the line for movement of nutrients and energy through

This bird's-eye view of the Adige Valley in Trentino, in winter,
shows the region's dense development and lack of connectivity
for wildlife. *Andrea Corradini*

the system. She kept studying the roe deer, but this larger research agenda was on her mind. When she first visited this crossroads of connectivity, she knew she had found the right location to study it.

"This is where I said, 'Okay, maybe now I found a place where I can put into practice my projects about ecosystemic relationships and how they have changed.'" She was trying to measure coexistence in the matrix, and also understand how nature was transforming as the climate warmed. How was habitat for wildlife shifting, and what did that say about the paths they would need to use, and how would those intersect, now and in the future, with human activity?

Cagnacci and her students and field assistants were spending the summer measuring things to try to get a handle on some of these dynamics. They had mapped out plots of a consistent size at sites around the region and were recording what was growing there, the height of the plants, how many were flowering or fruiting. At the edge of a steep meadow, with the sound of cowbells jangling around us from nearby herds, the team took samples of the soil to see what microbes lived inside and clipped bits of plants to measure their chemical signatures, a way to understand the flow of energy. They searched for poop—signs of presence!—to identify which animals had been passing through.

It seemed like a nice way to spend a summer, rooting around in a postcard-perfect Alpine meadow. But like any fieldwork, it involved long, repetitive days—ones that would hopefully, ultimately, yield important findings. Cagnacci had also helped launch another new project, the Global Initiative on Ungulate Migration, which links up research and conservation around the world so that scientists can share tracking data on the planet's hoofed mammals—elk, bighorn sheep, pronghorn, wildebeest—and find new ways to monitor and protect them.

Understanding these movement patterns is bigger than protecting the creatures themselves.

Animals grazing, browsing, stomping the ground with their
hooves, eating here and peeing over there, being taken by predators or dying another way and being eaten by scavengers such
as vultures: All these things influence ecosystems, directly and
indirectly. They move nutrients around the system, affect the
way carbon is stored in the soil, and can even change the likelihood and intensity of wildfires. Those same processes are
at work around the globe. Humans and large roaming mammals have shared landscapes for thousands of years; their
continued coexistence is essential—ecologically, energeti-
cally, and ethically—whether in the crowded, multiuse habitats of Western Europe, or the vast, big-sky landscapes of the
American West. There, coexistence has long relied on fencing—
but new ways of thinking were beginning to tear down some of
those walls.

Wyoming's Greater Yellowstone Ecosystem

A Fence Runs Through It: How Fences Make Us Bad Neighbors

West of Cody, Wyoming, along the road to Yellowstone National Park, the North Fork of the Shoshone River winds through the Absaroka Mountains, a landscape of extinct volcanoes that once towered thousands of feet higher than it does today. Strange formations of eroded volcanic rock, known as hoodoos, cap the hillsides. If you're lucky, you might see a flock of bighorn sheep scampering beneath these ancient deposits. But that's much less likely than it would have been before White colonists first began developing—and fencing—the landscape.

Ever since my first trip to the Rockies of the northern US in 2003, I'd been gobsmacked by the extent of the fencing there. I fell in love back then, as so many do, with the region's vast

---

Previous Spread: Bison graze during the rutting season in the Hayden Valley, Yellowstone National Park. Many of the charismatic wildlife species that draw visitors to Yellowstone move in and out of the park throughout the year, which makes conservation on nearby private lands crucial. Seemingly wide-open landscapes in the American West are often sliced up by miles and miles of fencing, making them difficult or impossible for some animals to cross. *Michael Nichols*

spaces, but as I drove around, I just couldn't get over the barbed wire. You can drive for the better part of a day, or longer, and have fence lines alongside you the whole way, wire barricades cleaving the seemingly boundless vistas.

Across the American West, after the Homestead Act of 1862 granted 160-acre tracts of land—stolen from Indigenous people— to pioneers willing to settle down and grow crops, erecting fences became an obsession. Fences were a muscular way to delineate property, and also a means to protect crops from livestock. But there weren't enough trees to build them from wood, and it quickly became clear that strands of wire did nothing to deter cattle. People began experimenting with ways to strengthen the wire with all sorts of sharp and pointy metal prickers, and in the early 1870s, a farmer named Joseph Glidden, who had moved west with his family from New York State to Illinois, invented a successful form of barbed wire that quickly became all the rage. Soon the stuff was being mass-produced, and nothing could stop pioneers from stringing it across their acreage.

Immediately, the barbed wire began causing problems for wildlife. Buffalo got trapped in the wire and either starved or died of injuries, not to mention that the fences kept them from reaching vital forage and watering holes. It's unclear just how big a role fences played in the near-eradication of buffalo across the West, compared to their mass slaughter for hides, but Native Americans, who had relied on the buffalo for centuries, nicknamed the wire "devil's rope." Within a matter of decades, of the massive herds of buffalo that had once dominated the West—there were 60 million of them, by some estimates—only a few hundred remained.

Today, scientists conservatively estimate that more than 600,000 miles of fences crisscross the West, and that's without counting property fencing in cities and suburbs. In just one Wyoming county, researchers mapped roughly 4,500 miles of

fences—that's nearly two and a half times the length of the US-Mexico border. The Absarokas and other Western landscapes may look wide open and sprawling, particularly to road-trippers from more paved-over parts of the country, but in fact they are sliced up by thousands of miles of barbed wire, put there to keep livestock in (or out), to mark boundaries between public and private lands, or to keep animals away from roads.

I'd come to the Absarokas to participate in a quiet but meaningful event, one that's slowly taking root in places around the world. On a dazzling summer day, I pulled off the highway and up a dirt road at the edge of a ranch, where a dozen or so people were already assembled in a field, gathering tools. From pickup trucks and SUVs, they retrieved wire cutters, thick gloves, buckets, and water bottles. They were ready for a morning of manual labor: The volunteers planned to dismantle several miles of barbed wire fencing. It was strung across private land, whose owners had actively sanctioned its removal—yet the enterprise still felt oddly subversive, almost like breaking and entering. Clipping a fence connotes trespassing; it was hard to get past that sensation. Perhaps it's a sign of just how entrenched the notion of fences as barriers has become in our collective psyche.

— — — —

Heading up to Cody, Wyoming, driving north across the state with my family, I marveled at how relatively empty the state's land still felt, compared to Colorado—where it's increasingly hard to go camping without a reservation, and nearly impossible to get one if you don't book it six months in advance. But while there may have been fewer people, the fences were everywhere.

West of Yellowstone National Park near the border between Idaho and Montana, barbed wire fending prevents wildlife movement. Fences often aren't visible in satellite imagery, which means they aren't considered in conservation planning. A new field of science called fence ecology is studying the impacts of fencing around the globe. *Karsten Heuer*

For hours and hours, they lined the highway, broken only where two roads intersected. We passed a deer standing frozen at a roadside fence. Was she trying to cross? Or did she not even bother anymore, having come to see the fence line instead as a territorial boundary?

Buffalo were nowhere near the only wildlife harmed by barbed wire. Fences can spell disaster for animals that migrate seasonally, or for those simply moving around for their basic needs. Mule deer can get their legs caught as they try to jump over. Pronghorn antelope, which tend to scramble under fences, can get stuck or scraped on the bottom wire, dying there or later from infected wounds. One study looked at pronghorn in the northern sagebrush steppe in Montana, Alberta, and Saskatchewan—where the research area contained enough fence to wrap around the planet eight times—and found that pronghorn selected paths with fewer fences to cross. The landscape might be vast, but the actual habitat available to animals can be far less than what it seems. This kind of information is crucial to restoring connectivity and permeability.

Because it's usually not possible to see fences in satellite imagery, they are often overlooked in conservation planning. They also do not factor into the Global Human Footprint Index, a widely used mapping tool that shows the relative influence of humans on landscapes around the world. In fact, said Wenjing Xu, a postdoctoral researcher at the Senckenberg Biodiversity and Climate Research Centre in Frankfurt, Germany, fences tend to proliferate precisely in areas that appear to have an otherwise low human footprint. She had noticed this same phenomenon the first time she traveled to the Tibetan Plateau, a landscape she had dreamed of visiting as a child in her native China. "From a human perspective, fences are for managing land and livestock, and they are barely visible from afar," she said. "For animals that need to roam, however, every 'invisible'

fence line could be an actual barrier that they have to figure out how to overcome."

No matter where you are across the western US, Xu said, on average you are likely less than two miles from a fence. As a PhD student at the University of California, Berkeley, Xu was part of an emerging field of research known as fence ecology, which has sprung up to uncover the impacts of all these fences—and find ways to mitigate them.

Despite being "one of the most widespread manmade features on Earth," according to one study, fences have never previously been the focus of comprehensive scientific research. This may be partly because, as that same study put it, fences are hard both to detect and to define. Is a border wall a fence? What about a line of beehives used to deter elephants? If you can't define something, it's tricky to study it.

Fences can be barriers to wildlife movement and cause ecological consequences that ripple out, but they can also be helpful tools for both ecological restoration and, somewhat counterintuitively, wildlife connectivity. Around the US and globally, the scores of wildlife overpasses and underpasses being constructed require miles of high fencing to keep animals off roads and funnel them to the bridges and tunnels designed for their use. Untangling the impacts of fences on all kinds of creatures that encounter them is a challenge. "Every fence has social and ecological winners and losers," said Christine Wilkinson, a postdoctoral researcher at the University of California, Santa Cruz, who studied wildlife and fences at Berkeley along with Xu. It can be hard to figure out which is which.

At Kenya's Lake Nakuru National Park, fences erected in the 1980s to keep poachers away from endangered rhinos have differing impacts on a range of species. Wilkinson set up wildlife cameras there for a year and captured more than 65,000 images of animals approaching the fence. Many animals were

able to cross—going under or over, sometimes digging their
way through—and did so frequently, including twenty-seven
mammal species: bat-eared foxes, spotted hyenas, even lions
and baboons. Carnivores and primates were best able to cross
the fence; ungulates, including giraffes and reedbucks, rarely
crossed. "Thinking about the other wildlife isn't something that
happens very frequently" when putting up fences for a specific
purpose, Wilkinson said. "We need to think about these non-
target species and ecosystem impacts."

Fences built specifically to manage one or a handful of species are generally successful at meeting their goal (whether to
keep a particular species in or out), but they often have terrible
unintended consequences. In Australia, where fences have long
been used to protect crops from non-native European rabbits,
and to protect livestock from native dingoes, in more recent
years land managers have built them to keep a slew of invasive
species—feral pigs, red foxes, domestic cats—out of ecological
restoration areas. In one case, at a roughly 2,000-acre reserve
in southeastern Australia, researchers built six-foot-high fencing to keep dogs, cats, rabbits, and foxes out of a restoration
project where they were reintroducing native and locally
threatened southern brown bandicoots and eastern bettongs,
two types of small marsupials. The fence succeeded in keeping
the invasive mammals out—but at a huge cost to local reptiles,
amphibians, and snakes, which died trying to cross the barrier
in one direction or another. Turtles overheated along the fence
line. Some turtles were even eaten by foxes right along the
fence, outside the enclosure.

A bearded dragon trapped in Australia's Dog Fence—a
common phenomenon. The Dog Fence, the longest fence on
the planet, stretches for more than 3,400 miles (5,600 km).
Built in the mid-twentieth century to keep dingoes away from
domestic sheep, it has been catastrophic for wildlife and
ecosystems. *Adam Ferguson*

The scientists who studied the fence's impacts counted thirty-seven other such "exclusion fence" projects around Australia, involving 526 square miles of land, plus more than 6,200 miles of dingo-proof, rabbit-proof, and emu-proof fencing around the country. None of those numbers include the untold miles of "undocumented" livestock fencing.

In Kenya's Maasai Mara grasslands, there's a different problem: An exponential increase in fencing is interfering with the huge seasonal migration of wildebeest and other animals. One study that documented the "unprecedented and accelerating speed and scale" of the Mara fences bore the distressing title, "Fencing bodes a rapid collapse of the unique Greater Mara ecosystem." When I asked George Wittemyer, a conservation biologist at Colorado State University who works with the Kenyan nonprofit Save the Elephants, about it, he was concerned. "It's a massive conversion of what was open plains, communal grazing lands, into privatized lots," he said. "It's going to be a huge area not accessible to wildlife. People are like, 'I'm going to fence off my lot so I can have dry-season grazing, and I don't want the wildebeest.'"

Elsewhere in Africa, in Botswana's Okavango Delta, "veterinary fences" built to protect cattle from wildlife diseases have blocked the passage of zebra, wildebeest, giraffes, buffalo, and other animals through a landlocked wetland. The fences have been linked to a decades-long drop in wildlife populations. Animals die trying to cross the fences or are picked off there by poachers. Over a six-year period of severe drought from 1979 to 1985, hundreds of thousands of migrating wildebeest died when the fencing blocked them from reaching a river. The Botswana

A guanaco cadaver hangs on a fence in Patagonia, Chile. Across the region, in Chile and Argentina, livestock fences like this one kill hundreds of guanaco, a type of llama, which get stuck on the barbed wire and starve. *Ingo Arndt/Minden Pictures*

fences even impact plants: Where the wildlife get stuck and become concentrated in a small area, plants die and woodlands erode into dirt.

You can find examples of the fence problem in virtually every country in the world. In Argentina's Patagonia region, pervasive livestock fences are death traps for guanaco, a type of wild llama, which get stuck on the barbed wire and starve. In just one two-year period on a single sheep ranch, 124 guanacos died this way. And, not to belabor the grim details, but in parts of Eastern Europe, there is more border fencing now than there was during the Cold War, and new fences designed to deter human migrants moving from the Middle East and Africa—a humanitarian disaster—are also carving up crucial habitat for carnivores like bear, lynx, and wolves. Closer to home along the US border with Mexico, the partially constructed wall, thirty feet high in some places, cuts off movement corridors for jaguars, ocelots, Mexican gray wolves, and many others, threatening more than ninety endangered species by one estimate.

As the changing climate increasingly forces animals to move to find what they need, all these fences around the world will become an even more dire problem.

— – – —

While some fences, like the US-Mexico border wall and those in the Mara, are new, others are relics, erected decades ago and no longer serving any purpose. In that way, taking down fences is similar to the movement to remove outdated, environmentally devastating dams along some rivers. As fence ecology raises awareness of the potential for harm, land managers and conservation groups are beginning to push for removal or replacement of fences—and it's sometimes a solution that otherwise-at-odds constituents can get behind. "Everyone can agree on this," Tony Mong, a wildlife biologist with the Wyoming Game and Fish Department, told me that morning on the ranch. Mong was the

chair of the Absaroka Fence Initiative (AFI), a group that orga-
nizes volunteer work days to take down old fencing that's need-
lessly blocking wildlife movement. (In places where fencing is
necessary, there are wildlife-friendly versions: Fences should
be clearly visible to mammals and birds, have bottom rungs
high enough for some animals to pass underneath, have top
rungs low enough for ungulates to jump over, and have enough
space between the top two rungs to stop deer from becoming
entangled. Some ranchers are now experimenting with "virtual
fencing," outfitting livestock with GPS collars that work with
invisible fence lines, beeping out warnings and sending small
electric shocks when the animals near the line.)

The Absarokas are part of the Greater Yellowstone
Ecosystem (GYE), a place where the migration of ungulates—
pronghorn, deer, elk, bighorn sheep—is "what makes this
whole system run," according to Arthur Middleton, a wildlife
ecologist who divides his time between Cody and Berkeley. The
migration here, much like the one in East Africa, is crucial for
the health of these animal herds. But it's also crucial for the
health of the ecosystem itself. Saving the planet's biodiversity
and stemming climate change are not separate issues. Another
emerging field of science, called zoogeochemistry, aims to
merge the too-often separate studies of wildlife ecology and
of the interactions between plants and soil—what scientists
call "biogeochemical cycling." Climate science relies on measurements of these plant-soil interactions because they deal
with how much carbon the planet's trees and other plants are
soaking up and how much is being stowed away, below ground,
rather than contributing to warming in the atmosphere. These
measurements, though, rarely include the role of animals, even
though their presence or absence can change the calculations
to a huge degree. Across the planet, animals contribute to storing billions of tons of carbon.

In the GYE, the ungulate migration is increasingly blocked by roads, pipelines, and residential subdivisions, as well as all the fences. "It's death by 5,000 cuts," Middleton said. The Absaroka Fence Initiative, though, was taking action. The group is made up of state agencies like Wyoming Game and Fish, as well as federal ones like the Bureau of Land Management, conservation groups like The Nature Conservancy (TNC), and local landowners, and had been organizing fence-removal and modification events for about a year—all during Covid times—when I visited. One event, in May 2021, drew about eighty volunteers, who removed three miles of fencing, nearly 4,000 pounds of wire and posts. The July effort was smaller, with twenty-five volunteers recruited from AFI's ranks rather than the general public.

We organized into teams and fanned out to different spots across the ranch. I tagged along with Xu and Middleton, as well as some people from TNC and the Mule Deer Foundation, and one of the ranch's owners. For three hours, we inched our way south and then east, between a road and an irrigated field, removing metal fasteners that attached the barbed-wire strands to posts. We clipped and rolled huge lengths of wire, untangling the strands from thick sagebrush that had grown up and entwined itself around the fencing.

Standing in the scorching sun of the late morning, the ranch owner, who did not want me to reveal her name or the property's, pointed out a well-worn elk path. "The mamas come through twice a day, morning and evening," she said. "They leave their calves in the sage and come down here. So they have to navigate the fence." She pointed toward another valley where there was

Greater Yellowstone Coalition staff retrofit a fence in Montana's Paradise Valley to make it safer for wildlife to traverse. Wildlife-friendly fixes can include raising bottom rungs and lowering top ones, and allowing enough space between rungs for animals such as deer to pass through.
Greater Yellowstone Coalition

a herd of twelve bighorn sheep. Moose were common here, too, and the previous week she and her husband had watched a herd of forty elk graze near their house. "We've wanted to take this fence down for years," she said. "But now with this new initiative starting up, it's much easier."

"Easy" was maybe relative, I thought. It was grueling work. As the sun rose higher, the temperature went with it. At some point, like a mirage, an ATV appeared—some other volunteers bringing us blissfully icy drinks. By the time the work was done for the day, we had removed two miles of fencing. It felt deeply satisfying, and also like a proverbial drop in the bucket. For volunteers, though, this kind of work yields instant gratification, a sense that you are having a true, direct impact on the land and wildlife. That connection is vital. I thought of the mama elk who now could access their browsing areas without having to clamber over barbed wire, and my own sweaty scramble in the sagebrush seemed more than worthwhile.

"There's a growing recognition of how important these working lands are," Abby Scott, Northwest Wyoming program director for TNC, told me over sandwiches at a spectacular hilltop cabin on the ranch. "I think we are right on the edge of becoming something bigger."

Middleton hoped so. About thirty miles to the west, Yellowstone is thronged with tourists year after year, many if not most looking to see some of the region's charismatic megafauna. But what lots of people don't realize is those animals don't stay inside the park. They move in and out across the boundary, and without the "out" part, they can't survive. "People don't make connections between the wildlife they want to see in Yellowstone and the conservation work that needs to happen here," Middleton said. It was similar to the issues playing out in Corcovado, in Costa Rica, albeit on a much larger scale: Some wildlife needs more than what a protected area alone can provide.

Middleton was working on ways to rebuild those lost connections, both physical and psychological. He had teamed up with the US Department of Agriculture, and they were testing a tool called habitat leasing, using existing programs in the Farm Bill to protect private working lands. The idea was that the government could rent a certain portion of land that the owner would maintain as wildlife habitat. Since the USDA first announced the plan, at a conference marking the 150th anniversary of Yellowstone National Park in the summer of 2022, the agency had budgeted more than $15 million in funding. Before a room packed with people who were deeply devoted to the Greater Yellowstone Ecosystem—conservationists, biologists, tribal representatives, ranchers, officials from multiple levels of government—Robert Bonnie, the USDA Under Secretary for Farm Production and Conservation at the time, talked about the limitations of regulations. "Federal regulations, for example, have little utility in stopping the conversion of open space for ranchettes, or helping ranchers install wildlife-friendly fencing ... or encouraging the restoration of habitat for use by the wildlife," he said. "In short, regulations can stop bad things, but they're often not very good—and sometimes they're even counterproductive—in encouraging stewardship."

Private lands comprise about 30 percent of the Greater Yellowstone Ecosystem, an area of roughly six million acres. So developing a new approach to private lands conservation was crucial. "Partnering with landowners to conserve habitats, dozens or even hundreds of miles away from the stone gates at Gardiner or West Yellowstone"—the park entrances—"may have as much

---

Next Spread: In Yellowstone National Park, grizzly bears dine on an elk carcass while wolves wait eagerly for a scrap. Animals move in and out of the park, and they need the "out" part to survive. *Dan Stahler/NPS*

bearing on these magnificent animals as what happens inside the park," Bonnie said.

Bonnie used to be a vice president at the Environmental Defense Fund, where he looked for ways to reward landowners for good stewardship. When we spoke about the habitat leases, he was optimistic. "We're taking these big iconic species in this big iconic ecosystem and recognizing, 'You know those elk you go to see in the summertime? Let me tell you where they are in the wintertime.' It asks a lot of questions that people in the environmental community have been able to avoid for a long time. It means we have to work with tribes, private landowners. That changes your perception of, 'What tools do we use?'"

The idea, he said, was to recognize that large mammals "bring some costs with them. They eat forage, they can be rough on fencing." Elk and moose "might drag a few grizzlies and wolves over the hills" in their pursuit. A flexible lease could be more appealing to landowners than a static, permanent conservation easement—a transaction that involves protecting an area from any future development. Conservation easements are appealing to some landowners but a nonstarter for others. The habitat leases are "a little bit a la carte," said Bonnie, and could be tailored to the specifics of a landowner's particular interests.

— – – —

Like other similar towns in the West, Cody was experiencing a post-Covid development boom. People from the coasts had been descending, people who could suddenly work from home and wanted to do it in beautiful scenery but were priced out of more built-up places like Jackson and even Bozeman. The newcomers didn't always understand the landscape. Much of the development pressure was happening right in the middle of pronghorn, elk, and deer migration routes, which is also key habitat for grizzlies. The tension in the town was palpable. I overheard locals at a downtown coffee shop disdaining

"Californians," which I took as code for all newcomers, and people everywhere were discussing the price of real estate.

Speaking on a panel about how to coordinate conservation efforts across the Greater Yellowstone Ecosystem, Albert Sommers, a rancher and state representative, talked about the huge pace of change in the region and the need to build relationships with the newcomers. "Figuring out how to reach this new breed of landowner" was crucial, he said, but "their aloofness is a key challenge." Sommers mentioned a billionaire landowner who he said built a golf course and a ten-foot-high, four-mile-long fence square in the middle of moose habitat and a mule deer migration corridor. "How do you reach these people?" he asked.

Another panelist mentioned a related issue. "Private lands are so important to so many species we care about," said Bob Budd of the Wyoming Wildlife and Natural Resource Trust, "but we are losing these lands because we send the wrong economic incentive." For landowners in Wyoming, he said, there were "two signals for the value of your land, wind and residential." Property owners looking to capitalize on their land's value could either lease it to wind farm developers or sell it off to be subdivided for housing. "There is no market signal for conservation."

Bonnie and Middleton hoped to change that. By the end of 2023, more than 300,000 acres of private lands had been voluntarily protected for at least a decade, through nearly a hundred individual contracts that paid ranchers eighteen dollars per acre and helped remove or replace fences harmful to wildlife and get rid of invasive grasses. By 2024, the program had been extended to Montana and Idaho.

There was no time to waste in promoting and expanding the conservation leases. As Middleton drove me back to town after our day cutting down the fences, we passed new homes going up on a subdivided former ranch on the valley floor. They were being built along a well-known wildlife corridor.

CHAPTER 7   *New York City & Long Island*

The Carnivore Among Us: Coyotes and the Urban Wilds

Seven years after Otis the coyote sparked a panic in uptown Manhattan, another coyote turned up in Central Park. Like Otis, this coyote, nicknamed Hal, was tracked, cornered, and darted, in a spectacle that involved an exhausting chase by the NYPD, the parks department, and of course the media. Hal was ultimately released into "a more coyote-friendly habitat," wherever that was. Again, officials could only guess how he had arrived in New York; they ventured that he might have crossed a railroad bridge, or else he swam beneath it.

All these years later, when mulling our increasingly chopped-up world and the way animals are moving through it, I think back to those two Central Park coyotes. There are so many losers in the era of fragmentation, but are there winners, too? Were the

---

Previous Spread: A coyote roams near the Whitestone Bridge in the Bronx, New York, one of three bridges that span the East River and connect the borough of Queens and Long Island with the mainland. Several parks in the Bronx now have permanent coyote populations. *Ivan Kuraev*

coyotes of turn-of-the-millennium New York a harbinger of a new coyote-friendly world?

While so many animals' habitats are shrinking, a few species are actually expanding their ranges—managing to thrive in spite of, or even because of, human activity. Coyotes are one of them. Back in 2004, researchers compared the contemporary and historical ranges for forty-three North American species of carnivores (foxes, badgers, lynx) and ungulates (pronghorn, moose, mountain goat). They found that, in general, both the ungulates and the large carnivores had lost a lot of habitat, with bears, wolves, cougars, and wolverines losing significant chunks of their ranges. But raccoons and coyotes, by contrast, were thriving. Raccoons had expanded their territory by nearly 20 percent, and coyotes by an amazing 40 percent.

Scientists argue about the exact historic ranges of coyotes, but the most recent research—based on museum specimens, records of sightings, and fossils, bones, and other remains found during archeological excavations—shows that from 10,000 years ago through the late 1800s, coyotes lived across about two-thirds of the current United States, from California to the Mississippi River. They also likely ranged as far south as northern Costa Rica, and north into southern Canada. But they had never been found in the eastern US; there was no trace of them in New York. In each decade since the 1900s, though, coyotes gained new territory: east across the US, south to Panama, and northwest across Canada all the way up into Alaska.

This continent-spanning coyote explosion was made possible by the very logging and fragmentation of forests that has made life so hard for other species, and by European settlers' wholesale slaughter of wolves. Wolves outcompete coyotes for food and will even kill them in the competition. (Research in Yellowstone and Grand Teton National Parks in the early 2000s showed that when wolves were reintroduced to Yellowstone,

coyote densities dropped by nearly 40 percent.) With the top-level predators missing from the food chain, coyotes flourished. The extermination of mountain lions across eastern North America also benefited coyotes, as did the decline of both mountain lions and jaguars in Central America.

The coyotes found east of the Mississippi today are a recent creation. The eastern coyote, which can now be found from Florida to Newfoundland, is actually a hybrid of coyote, wolf, and domestic dog, with those in the northeast possessing the highest amount of wolf and dog DNA—between 8 and 25 percent wolf and around 10 percent dog, according to research by Roland Kays, a zoologist at North Carolina State University. Kays estimates that these coyote-wolf mating events happened around a hundred years ago. "A century ago, wolf populations in the Great Lakes were at their nadir, living at such low density that some reproductive animals probably couldn't find another wolf mate, and had to settle with a coyote," Kays has written. The wolf genes made the coyote slightly larger, which likely gave it the ability to make dinner out of deer—a ready source of food in eastern forests, which allowed the hybrid coyotes to expand their historic range. "These animals thrived, dispersed east, and thrived again, becoming the eastern coyote." The dog mating was more recent—roughly fifty years ago, by Kays's estimate. "Nowadays," he wrote, "eastern coyotes have no problem finding a coyote mate. Their populations continue to grow throughout their new forested range, and they seem more likely to kill a dog than breed with it."

Unlike some other critters that have colonized new areas in the last century or so, like emerald ash borers or feral swine,

---

An eastern coyote in Massachusetts. Eastern coyotes are a hybrid of coyote, wolf, and domestic dog. The wolf genes make this subspecies slightly larger, which probably helped it thrive as it continued its eastward march. *Ivan Kuraev*

both of which were brought by humans and caused havoc to ecosystems, coyotes evolved and adapted to *exploit* a human-altered landscape, taking advantage of new opportunities and then using their wits to survive. In the 1980s, they made it as far south as southern Panama, and now only the dense forests of the Darién Gap separate them from Colombia and South America, where they have never existed before.

Curiously, a South American midsize canid is making the opposite journey: The crab-eating fox, short-legged and bushy-tailed with a thick coat of fur, crossed the continental threshold from Colombia into the Darién Gap perhaps a decade after coyotes first appeared in Panama. As one animal moves north and the other south, the coyote and the crab-eating fox are now sharing habitat for the first time ever in the area between the Panama Canal, the northernmost extent of the fox's expanded range, and the Darién Gap. No one is quite sure yet whether or how they are interacting, but an extensive network of camera traps has confirmed the presence of both animals. It's the first time predators have swapped American continents in three million years.

Kays thinks the coyotes' movement through the dense forest of southern Panama will be aided by a human presence there: the thousands of migrants who risk their lives to journey north, and drug smugglers who use the dense forest as a hideout. The people, Kays said, "are probably scaring away the jaguars, probably leaving scraps of food, and stuff like that."

Some scientists are excited about these carnivores on the move. Some, less so. When I mentioned coyotes to a Belizean conservation biologist, she bristled. "We have started to see coyotes in the forests here, and it is a very bad thing," she said. She worried that they would gobble up small animals of all kinds, putting species in peril and ecosystems out of sync.

But Kays disagrees. "I think there's a lot worse environmental catastrophes out there than coyotes," he told me.

"The places where predators have the biggest impact is when the prey have never seen a predator like that before," Kays said. But in Central and South America, other native midsize carnivores, like foxes and coatis, as well as domesticated dogs, occupy the same ecological niche. (Recent photographic evidence shows that Central American coyotes are continuing to breed with dogs, which could also be aiding their continued expansion.) Kays believes we should be more concerned about the conditions that allowed the coyotes to expand into so much new territory and have hurt so many less adaptable species: deforestation, roads, the demise of top-level predators— ecosystems decimated and already out of balance.

— – – —

Coyotes are generalists, meaning they aren't picky about either food or lodging. This has helped them expand from grasslands into forests, suburbs, and cities. But they also benefit from their particular relationship to humans. Scientists use the term *synanthropic* to refer to a type of animal that thrives in urban landscapes—rats and pigeons, geese and foxes, raccoons and coyotes. A synanthropic species might actively choose to hang out in human-dominated places, moving into a new range, or it might simply be better equipped to survive in cities than other animals. Species that prefer instead to avoid humans, or that are harmed by urbanization, are called *misanthropic*.

The vast majority of carnivores that live on land are misanthropes; fewer than 15 percent are synanthropes. But scientists who have spent more than twenty years studying coyotes in the Chicago metro area believe that coyotes are a strange mix of both. In one study, those researchers put radio collars on 181 coyotes over six years and wound up, they wrote, with "a portrait of an animal that appears to benefit from the urban landscape through enhanced survival and possibly elevated population densities, while also exhibiting strong spatial and

temporal avoidance of humans." The animals were avoiding more developed areas of the city and also becoming nocturnal. It's a wily strategy: thriving in busy cities while steering completely clear of people. This status, which the scientists referred to as "synanthropic misanthropes," helps both individual animals and the species overall.

The same is true in New York City. Coyotes are thriving there—and scientists are on to them. Since 2009, the Gotham Coyote Project (GCP), a collaboration of researchers, has been studying the ecology of the city's coyotes. I called up Anthony Caragiulo, a geneticist who worked on the project and also served as assistant director of genomic operations at the American Museum of Natural History's Institute for Comparative Genomics.

For Caragiulo's PhD work, he used mountain lion scat to study the genetic relationships between wild cats that lived throughout Central and South America. In order to study the scat, someone first has to locate and then "collect" it—in the mountains and jungles and other wild places where cougars live. Caragiulo, though, was not that guy. He had chosen to work in the relative safety of the lab. "I don't need to go out in the field with botflies!" he joked, adding that he doesn't even like to go camping.

For the Gotham Coyote Project, teams collected coyote scat closer to home. Caragiulo was analyzing DNA from coyote poop collected in parks around the city and suburbs. But he once again stayed in the lab, waiting for the scat to come to him. "I don't go out and collect anything," he said when we spoke on the phone. "People just send me boxes of crap in the mail."

A radio-collared juvenile male coyote in Chicago, where the Urban Coyote Research Program has studied the animals since 2000. This coyote, who roamed across the city to find a territory for himself, sadly was hit by a car just a few months after the photo was taken. *Corey Arnold*

Coyotes, it turned out, had established permanent residence in the city's outer boroughs—but not, at least as of yet, in Manhattan. In January of 2020, though, as Covid-19 cases began popping up in the US, another coyote passed through Central Park. By then, New Yorkers had grown more blasé; instead of a chase, there was a tweet. The NYPD told people to "observe and appreciate coyotes from a distance."

Like many New Yorkers, coyotes were paying occasional visits to Manhattan but actually resided in the outer boroughs. In fact, they were thriving in the Bronx and continuing to expand their territory, moving across the East River into Queens and out toward the suburbs and rural areas of eastern Long Island. The researchers were trying to build a "genetic connectivity map" of all the coyotes, to understand who was related to whom, and where they had come from. Caragiulo's role in building the map was to extract the DNA from coyote poop he received in the mail. It was hard for me to picture: post office poop. "So, do people just, like, put it in a box and write your address on it?" I asked him.

"They put it in a brown paper bag," he answered. "And sometimes they throw in some silica beads, like the ones that come when you buy a pair of shoes."

"And it just shows up in your mailbox at the museum?" I had spent a good part of my childhood in that building, and I still viewed it with the awe and reverence of a kid. If you'd asked me what I thought might arrive in the mailboxes of the museum's venerable scientists, I'd have said exotic gemstones and the perfectly preserved vertebra of plesiosaurs. Definitely not coyote poop from the Bronx.

A coyote in Ferry Point Park, at the foot of the Whitestone
Bridge in the Bronx. Coyotes are expanding their range,
colonizing a landscape where they have never existed before.
Ivan Kuraev

Caragiulo graciously invited me to see his inbox for myself, so one morning I met him at the museum's basement entrance. It was a Tuesday in the fall of 2021, and because of pandemic-related staffing shortages, the museum was closed on Tuesdays. Other than security guards and some crews working on exhibit maintenance—and a public vaccination clinic underway in the lower-level lobby—the building was eerily empty. We headed to a giant elevator and rode it up to the eighth floor. The hallways were deserted, and it felt vaguely apocalyptic. Because of the pandemic, most people were still working from home, he said, only coming in when necessary and allowed. The pandemic had also caused supply chain issues in the lab. Normally abundant items like pipette tips and DNA sequencing kits were taking up to three months to arrive. This had slowed down Caragiulo's coyote work. The poop was piling up in his lab. "I'm the bottle-neck," he said.

He walked me over to an area of the lab where the shelves and surfaces were lined with cardboard boxes and precarious stacks of envelopes. At a station marked "Bench 3—scat bench," he opened a box that in turn was filled with small paper bags, lots of them. "Cedars Golf Course, Cutchogue," read the hand-written info on one bag, from a scat collector named J. Murray, picked up on January 8. "Near goose carcass and possible vomit of goose feather & bones," the notes stated.

I could see why Caragiulo preferred the lab. Other bags had different people's names on them—some scientists, some volunteers—as well as the names of various green places throughout Long Island: Quogue Wildlife Refuge, Kings Point Park. All had GPS coordinates.

Caragiulo donned a pair of blue latex gloves, spread a piece of tissue on the table, and opened up a paper bag. He retrieved a small bit of dried scat. It had no noticeable smell, and it looked like nothing more than a clump of mud. He shrugged.

"This could be anything. I have no idea." Most of the samples he had analyzed over the past year, though, were definitively coyote. (A few domestic dogs, the odd fox.) By extracting and analyzing the DNA from the scat, he could quickly tell, first, whether it did in fact come from a coyote. Then, using so-called microsatellites—short, repeated segments of genetic code that are unique to an individual—he could examine which samples came from the same individual coyote, and which came from coyotes that are related. He could also determine just how closely related they are. "The goal is to say, over five, ten, fifteen years, which families were successful, where they colonized— to look at this web of movements that's a snapshot in time."

Carol Henger, a molecular biologist who studied coyote DNA for her PhD research, said each park in the Bronx had its own distinct family group. The DNA from the scat of each individual and each family allowed you to see just who was moving where—like, "this coyote at New York Botanical Gardens is also related to Pelham Bay coyotes," she said. "There was dispersal all around the city." From that genetic connectivity map, Henger and the Gotham team were also able to hypothesize how coyotes settled in New York City in the first place. The likely tale was that "they first settled in parks to the north of the city"—moving down from the suburbs of Westchester— "and their descendants have since colonized other parks and have their own family groups. The coyotes in Queens"—the newer arrivals—"were related to the coyotes in the Bronx." It was, like many New York stories, a tale of immigration and survival.

Scientists are interested in this particular snapshot because they can now watch as coyotes move into a landscape where they've never existed before: Long Island, the 120-mile-long, vaguely crocodile-shaped spit of land that juts into the Atlantic Ocean east of Manhattan, anchored by Brooklyn and Queens at

the western end. "We say things like, 'Oh, black bears are coming back, or bobcats are coming back,'" Chris Nagy, a cofounder of the Gotham Coyote Project, said, speaking about species whose populations are rebounding in places where they used to exist before humans caused their numbers to plummet. "But coyotes are showing up for the first time." Nagy marvels at this. "When I talk to people about this stuff, I say the city thing is cool, but they're also in Alaska all the way to the tropical rainforest in Central America. They're figuring out ways to make a living wherever they can."

As Kays put it, the coyotes "didn't start in the cities. They started in a better habitat. Then as they filled those habitats up, the juveniles started moving more and more into urban areas and suburban areas." He had seen it in his own neighborhood in Raleigh in the past couple of years. "All of a sudden, there were a lot of 'Lost Cat' signs, and I got more coyotes on my camera trap in my backyard. They've been here since the '90s. But any coyotes in the suburban areas were probably just kind of passing through, not breeding. But we definitely have breeders now in our neighborhood."

Whether you are celebrating the rare success story involving a carnivore or worrying that, like a plague of locusts, the animals might presage some ecological cataclysm, from a scientific perspective, watching a species colonize a new area is an uncommon opportunity—which makes it kind of cool. "It's really rare that we get to see the front of a species moving into new territory," Caragiulo said. The Gotham Coyote Project researchers tracking the coyotes "are only a year or two behind their movement, tracking the founding population. If, in thirty, forty, a hundred years from now, people are still here, and Long Island is still here, and there are still coyotes, we can see how they have changed from those original individuals." They will know in part by studying the desiccated evidence

that may still be arriving at the American Museum of Natural History's mailroom.

— – – —

Caragiulo notwithstanding, most coyote researchers do go into the field. Which is how I ended up, on a blustery October day, standing on a dead-end street between two used car dealerships just off the Northern Parkway in Queens, waiting for Nagy, the Gotham Coyote Project biologist who tracks coyotes around New York City. As gusts whipped the rows of American flags that ringed one of the car lots, I turned my back to the wind. A metal barricade separated the cul-de-sac's end from the edge of Alley Pond Park, 635 acres of green space stretching south from the edge of Little Neck Bay, across the water from the Bronx. From this vantage point, it looked mucky and uninviting. A dumpster sat where the sidewalk ended, next to a "No Parking" sign. The previous day, a storm had dumped several inches of rain on the city, and now orange traffic cones sat half submerged in a giant pond-like puddle, whose surface rippled in the gale. What on Earth was I doing here?

I stood on that desolate street for nearly half an hour, waiting for Nagy, wishing I had worn an extra layer. Each time a car turned off the highway I wondered if it would be him. I had no idea what kind of car he drove, but I was betting on a Subaru. Finally, I got a text. "How's it going? On your way?" I was confused. I was waiting for Nagy, but he was seemingly waiting for me. I walked around the dumpster and discovered a green Subaru that had been parked there the whole time, obscured by the giant trash bin. Sometimes a barrier is highly context-specific.

Nagy had short dark hair and wire-rimmed glasses and a boyish face. He looked a bit like Harry Potter transfigured into a field biologist. In addition to running the coyote project, he's the director of research and education at the Mianus River

THE CARNIVORE AMONG US

Gorge, a nature preserve in Bedford, about an hour outside the city. He and the coyote project's other cofounders—Mark Weckel, who also works at the American Museum of Natural History, and Anne Toomey of Pace University—"got into this in our free time," he said. Because NYC officials wouldn't give them permits at first, they began by setting out camera traps in Westchester County, just outside of the city.

Ultimately, the city relented, and Nagy now monitors fifteen parks around the city, using camera traps to piece together a snapshot of where coyotes are living and breeding. "For years and years we have been putting out cameras at certain sites— and watching them fill up with coyotes," he said. Nagy rattled off a list of parks in the Bronx that now have permanent coyote populations: Van Cortlandt Park, Pelham Bay Park, Ferry Point Park. According to anecdotal evidence and Nagy's cameras, the Bronx coyotes arrived first, simply walking from Westchester. They first appeared in two big parks right on the border. Next, they showed up in Bronx Park, in the borough's center. Then they arrived in a couple of parks on the south coast, along the East River. There's one male coyote who has successfully produced offspring almost every year since 2012. "We think it's that guy who's put out so many babies that fueled the dispersers that have now made it to Queens and Nassau"—on western Long Island— Nagy said, sounding almost proud. "I'm excited to see the genetics of the scat we found there so I can see who's the grandpa."

It's easy to understand how coyotes arrived in the Bronx, since the borough is just the southern tip of the same land mass as Westchester and the rest of New York State. But Queens is part of Long Island. How were they getting there?

Biologist Chris Nagy, a founder of the Gotham Coyote Project, sets up a camera trap near NYC's Whitestone Bridge—in the exact same location as the coyote on pages 196–97. *Ivan Kuraev*

"If you look for green on a map of the city, and see how that's connected, you can have a best guess of how they move through," Nagy said. When moving to new territory involves crossing water, he looks for "the shortest jumps to get where they want to go." The bridges from the Bronx to Queens that cross the East River—the strait that also lines the east side of Manhattan, separating it from Brooklyn and Queens—have no walkways, and in any case they are busy with cars all night long. They also involve long, curving on-ramps, which wouldn't make intuitive sense for an animal trying to move in a specific direction. Nagy and others believe the coyotes are swimming rather than walking—paddling from one small island to another across to Queens.

When I first spoke to Nagy, the swimming was just a best guess. Coyotes had been spotted on Randalls and Rikers Islands, two small islands (one a sports complex, one a prison) in the East River. But then, in 2023, the NYPD fished a coyote out of the river. Evidence appeared on Instagram: a video of the police department's Harbor team hauling the struggling animal to the safety of a boat. The soggy female was brought to a city animal care facility, which noted that she would be "safely released back into the wild." I couldn't help but wonder if perhaps she'd end up with Otis's pack of misfits at the zoo.

Our task on the windy day in Queens was to retrieve a series of camera traps Nagy had set out several months earlier. Along for the ride was his nine-year-old elkhound, Ethan, who bore a passing resemblance to a coyote—shaggy fur, bushy tail, pointy ears, multicolor coat—enough, at least, to help me imagine how they might look in this landscape. The three of us stepped off the pavement and into the park. We tromped across a wooded area, autumn leaves crunching under our feet, following muddy tire tracks—and the blaring sound of chainsaws. Park maintenance crews were doing restoration work, removing invasive

vines. Nagy had last visited the park in June, to place the cameras. "A year ago this was totally dense, full of thorns," he said.

Nagy followed his GPS readings to find the camera, which was attached to a tree in a small clearing, perhaps a foot and a half off the ground, secured with a wire cable and locked. He knelt down and removed it from the tree and opened the casing—and an entire ecosystem of bugs came skittering out. He laughed, turning the case upside down and tapping it against his hand.

Camera traps, by their nature, provide limited data. You can't use them to count animals, for one thing, because they only show the creatures that pass that specific spot. And while they can provide definitive evidence that a particular type of animal is present in the area, they can't determine whether an animal is absent. It might be in the area but the camera simply never gets a picture of it. Setting cameras in likely areas of passage— near water sources or on well-worn game trails—is crucial. If you put a camera in the wrong place, you may see nothing, even when there are ample creatures passing by. "If I put one on that side of the tree, all the animals that walk on this side I'll never get," Nagy said.

Placing game cameras involves calculations of probability. Raccoons are common in Alley Pond Park; if you left four cameras in four different spots for four months, you'd probably capture them in at least a couple of shots. But Nagy had had three or four cameras in the park every year since 2011, and the previous year had been the first time any coyotes had appeared in images—meaning, most likely, that the animals had only recently arrived.

We followed a dirt path along one edge of the park, where a steep slope rose up to a street of large, single-family houses. On the other side of the path was a muddy ditch, along which Ethan was happily loping, dodging between the water and the bushes

nearby. If I was a coyote, I thought, this was definitely a path I'd use often. I asked Nagy why he didn't simply set a camera in one of the trees that lined the canal. "I'd love to. But if I put a camera here, it'd get stolen. But yeah, for sure, animals use this path." Urban areas present another level of challenge. "I'm looking for a place that won't get found by people, that's number one."

Nagy thinks that NYC's coyotes follow the coastlines until they find a green space. "Coyotes have a really strong dispersal instinct," he said. "They can disperse hundreds of miles over open prairie. The raw distances in New York are nothing. It's the configuration of people and cars and roads" that pose the most danger for coyotes. The city's coastlines "are not beaches and public spaces" where people congregate and recreate. "In the Bronx they are these industrial zones, with reinforced sea walls—so at night people aren't on them." In other cities, like Chicago, coyotes use train tracks.

A few years ago, a team managed to catch and radio collar a family of coyotes living in a small park on the water in the southern end of the Bronx. There were two more small parks nearby, where local residents had also spotted coyotes. "We found that in these three parks along the water it was this one family running back and forth," Nagy said. From one end of the chain of parks to the other was a distance of roughly three kilometers.

"If they could use just the one park, they would've"—to avoid leaving the green space—"but they used three," said Nagy. With help from a high school student (GCP likes to involve students whenever possible) and using other coyote research as a guide,

---

A curious coyote investigates one of the Gotham Coyote Project's wildlife cameras, in the Bronx. Many urban coyotes utilize parks; they have learned to share spaces with humans who keep relatively predictable schedules, as long as they have places to hide when the people are out and about. *Christopher Nagy*

Nagy concluded that urban coyotes need at least a third of a mile, or one square kilometer, of green space within a five-square-mile territory to be able to survive and safely navigate the treacherous urban environments between sheltered islands of green. That would be about a third of the size of Central Park, in an area roughly the size of the southern end of Manhattan, below 23rd Street. The biggest park in that area, Washington Square Park, is just four-hundredths of a kilometer in size, which may be why only one coyote has been spotted in southern Manhattan. In 2016, a coyote wound up in Battery Park City, at the southwestern end of the island; police relocated him to the Bronx.

"I ran that window across the whole city and got a map of where that could be," Nagy said. "Including little parks that maybe added together could suffice." Of course, the map is just theoretical. "You could have a place like Central Park that has great landscapes but 100,000 people a day, and that plays a role, too." But it gives Nagy, at very least, a good idea of where he might want to set out camera traps.

Even a developed urban setting can be a coyote stomping ground, as long as there are predictable times of day without people. Like construction areas, say, where the crews go home at five o'clock. Nagy found coyotes living in a construction site in Queens, where a patch of forest had been bulldozed to build a new parking lot for LaGuardia Airport employees. The construction started in 2014, and the coyotes had the place to themselves after the crews left in the evening. The coyotes only left after the lot finally opened in 2016. The construction cycle had been predictable, with workers leaving each evening—but now airport staff were coming and going around the clock. Nagy has observed the same thing on a golf course in the Bronx. "As long as there's somewhere to hide, they don't mind. There are people there but they never go into the bushes. So you have lots

of people, activity, noise, but it's predictable. It's a pattern the animals can learn and count on."

Following his GPS, Nagy ducked off the trail to the left, into a wide wooded area between the path and the steep berm, and found a camera he had placed in an area that felt both secluded and open enough to afford a view of the territory. Ethan sniffed around nearby, and it was easy to imagine a coyote loping through here. We returned to the car and drove around to the other side of the park to retrieve a third camera. This section of the park contained a forest with trees taller than any I have ever seen in New York City, some of their leaves beginning to turn yellow and others already on the ground. I could barely see Ethan in the low scrubby bushes that dominated the understory. He was camouflaged, except for his fluffy tail.

Nagy dropped me at LaGuardia on the way back from the park. Driving past the belt of parking lots around the airport, I thought about the coyotes that had once taken up residence there, and how much their fate depended on human actions, attitudes, and whims.

It didn't take Nagy long, after he began sifting through the Alley Pond Park footage a few days later, to find a coyote pic, taken at 11:58 p.m. one October night, with the animal about to climb over a fallen log. It was just one of many city critters the camera caught: a hawk perched on the log eating a small rodent, raccoons, rabbits, cats.

Even the biggest, densest, concrete jungle of a city is wilder than many of its residents imagine. Humans have reshaped landscapes, and pathways across them, in every corner of the planet—but no more so than in cities, where we've made it impossible for many animals to pass through. But not all.

A year after my visit to Caragiulo's lab, he published a paper revealing the genomic makeup of sixteen NYC coyotes. All contained dog ancestry—not surprising, given Kays's finding that

northeastern coyotes were hybrids. But one male turned out to be a "first-generation coyote-dog hybrid," with 46 percent dog DNA; two other coyotes were his offspring, and they each contained a quarter dog DNA. The three also carried genetic variations "known to influence human-directed social behavior in canines," whereas none of the other thirteen coyotes did.

The three coyote-dog hybrids were part of a pack that had lived near LaGuardia and was euthanized in 2016; concern that the animals posed a danger to people led transportation officials to call in the USDA to kill the coyotes. Previous Gotham Coyote Project research showed that those coyotes were eating a diet that included rats as well as "anthropogenic remains"—plastic and foil wrappers. "These coyotes faced a difficult existence," Caragiulo wrote, "as their genetics may have predisposed them to human interactions in an environment where such interactions do not typically have a positive outcome for wildlife."

Urbanization has changed the way animals move and access the resources they need to survive, creating clear winners and losers: Some species simply cannot cross a city, while others adapt and evolve to live among humans, one way or another. But that adaptation provokes its own set of transformations, physically and philosophically. It creates, on some level, a new and different form of wildness. In addition to being shaped by wolves and dogs, these urban coyotes are also deeply shaped by humans: by our past actions that enabled their continental spread, by our current responses to their growing presence in our built environments, by the way we live daily—our garbage, our social rhythms, even how we treat one another.

On the other side of the country, in San Francisco, a coyote nicknamed Carl became a beloved city fixture, celebrated for being "unusually friendly" and wandering through urban parks in broad daylight. He amassed devoted fans and a Facebook page, but some people worried that he was getting too bold. There

were occasional reports of him getting too close to toddlers. In 2021, he was caught on video heading across a busy park, straight toward a small child who sat on a wall fishing with his father; the man snatched the child away in what seemed like the nick of time. Officials decided they had to act, and not long after that incident, they fatally shot Carl.

To Christine Wilkinson, the ecologist who studied fences while at Berkeley and also researches coyotes in the Bay Area, Carl's tragic story shows yet another type of connectivity. "The likely reason he got so bold is because he used to live in a neighboring park and was fed by a homeless woman for many years," Wilkinson said. "That woman was a victim of so many societal injustices, and it trickled down to Carl." Wilkinson's coyote work combines information about wildlife movement with data normally used for more human-oriented research, such as neighborhood socioeconomic indicators and air pollution measurements. "I'm trying to bring together historically siloed datasets about human and wildlife wellbeing," she said. "Air pollution, water pollution—we know those things are bad, for humans and wildlife. I think there are a lot of intricate connections that we don't see." How these synanthropic creatures live, and sometimes thrive, in cities reveals things not only about animals and their needs and life stories and ecology, but also about us, and the social, economic, and political forces that shape our shared environments.

CHAPTER 8   *Los Angeles & North Carolina*

Corridors of Injustice: How We Treat Each Other Is How We Treat the World

All around the world, social inequality and loss of biodiversity are closely intertwined. The more uneven the income distribution in a society, the greater the decline in its biodiversity. A study published in 2007 compared a measure of wealth inequality called the Gini ratio to data on threatened species and found "striking relationships between income inequality and biodiversity loss." As we do to one another, so we do to the planet.

Within societies, and even within cities, a corollary relationship exists: Wealthier places tend to have more green space, with more vegetation providing more habitat for more species, than poorer ones. This trend can start at the individual backyard level and scale up to neighborhoods, and it also holds true at the city scale: Wealthier cities have bigger and better networks

---

Previous Spread: In Southern California, oil production infrastructure sits adjacent to a lower-income residential area. Patterns of income inequality—which are often tied to systemic racism—influence urban greenery, which also impacts the distribution of wildlife in cities. *Johnny Miller*

of urban parks. That, in turn, can mean more well-connected habitat for birds, bugs, and mammals, which impacts things like how many of those creatures there are, how healthy their populations are, and whether those populations are breeding and exchanging DNA with other nearby groups.

Scientists documenting this phenomenon in the sprawling metropolitan area of Phoenix, Arizona, more than twenty years ago termed it "the luxury effect." They set out to understand what factors might be influencing how many different types of plants occurred in a particular area. Was it determined by the altitude of a neighborhood, say, or whether it had previously been farmland? The answer, it turned out, was money. In the area they surveyed, the median family income was just over $50,000 a year. But in neighborhoods where incomes were higher, the scientists found that there was, on average, twice as much plant diversity.

Since then, scientists have found and documented the luxury effect all over the world. More recently, a widening group of ecologists, including Christine Wilkinson, has begun unpacking the structural underpinnings of the luxury effect, uncovering the ways that our own systems of oppression and injustice leave their mark on the natural world. They are asking questions like: How does systemic racism impact ecosystem health? How did—and do—social injustices perpetuated over decades and centuries influence the landscapes of cities, and how does that in turn shape the fate of both humans and wildlife?

"Often the interests of other species and marginalized humans align," said Madhusudan Katti, an ecologist at North Carolina State University, whose work considers the intertwined lives and fortunes of urban-dwelling species, human and animal. "If you're going to think about reconnecting habitats, it's useful to think about people who live there, not just wildlife." He added, "It's very much a colonial perspective to

think about humans and wildlife as separate. We need to start thinking about humans and wildlife together in the landscape, and mitigate things that will help both."

— – – —

In the fall of 2020, as the Covid pandemic laid bare the deep and enduring structural racism in US society and Black Lives Matter protests spread across the country, Christopher Schell, an ecologist at the University of California, Berkeley, published an influential paper in the journal *Science* that explicitly connected the dots between racism and the health and fate of non-human species. Titled "The ecological and evolutionary consequences of systemic racism in urban environments," it synthesized and brought into the zeitgeist what a handful of urban ecologists were already learning: that the luxury effect has a racial component, and in fact race can be a greater socio-ecological determinant than wealth; and that as a result, patterns of bigotry and oppression impact how other species experience our cities.

Schell, who is Black and from Los Angeles, said he grew up with an understanding that "there is a ton of heterogeneity that exists in a city, and it's not by accident that it's that way." Those variations could include the numbers of parks and street trees in different neighborhoods, whether a highway or rail line ripped through a community, or whether an oil refinery spewed toxins into the air. He wanted to show that urban heterogeneity in turn "is driven by systemic inequities like oppression, residential segregation, gentrification and displacement, unjust zoning laws, homelessness, so on, so on, so on." Those issues don't only impact people, he added: "How we operate influences the rest of the natural world as well as the social world."

In the paper, Schell and his colleagues argued that the uneven distributions of plants and animals throughout urban areas are directly tied to racist policies and attitudes, and that therefore structural racism has evolutionary impacts for wildlife, which

can be harmed by the same negative factors that harm humans. An unending sea of asphalt, for instance—or what scientists call "impervious surface"—can lead to "reduced movement of organisms across landscapes and therefore lower gene flow" for wild animals. Lower gene flow leads to reduced genetic diversity, and a species or population with low genetic diversity is at far greater risk of extinction.

Previously, environmental justice and urban conservation had tended to stay in their own corners; after all, one was about humans and the other about wildlife. But the Schell paper made clear that these two issues are inseparable. Cities are ecosystems, and they need to be healthy and functioning in order for all their inhabitants to thrive. "The insidious white supremacist structures that perpetuate racism throughout society compromise both public and environmental health," the paper concluded.

In Schell's lab, he told me, researchers "oftentimes do our own version of 'six degrees of separation from Kevin Bacon,'" to show how human actions ripple out to wildlife. "Air pollution isn't just restricted to people," he said. "Other animals have lungs. Why would we not expect them to also be inhaling the same amount of pollutants that we generate?"

— — — —

This growing field of inclusive urban ecology has been aided, in the US, by a tool called Mapping Inequality, a sprawling, multi-university project spearheaded by the University of Richmond's Digital Scholarship Lab. It created a digital archive of "redlining," the New Deal–era housing policy that enforced and perpetuated neighborhood segregation in the United States.

In 1933, the federal government created the Home Owners' Loan Corporation, or HOLC, whose intent was to help Americans recover from the Depression. HOLC, which was the predecessor to today's government-backed mortgage system, issued accessible home loans or refinanced mortgages to prevent default.

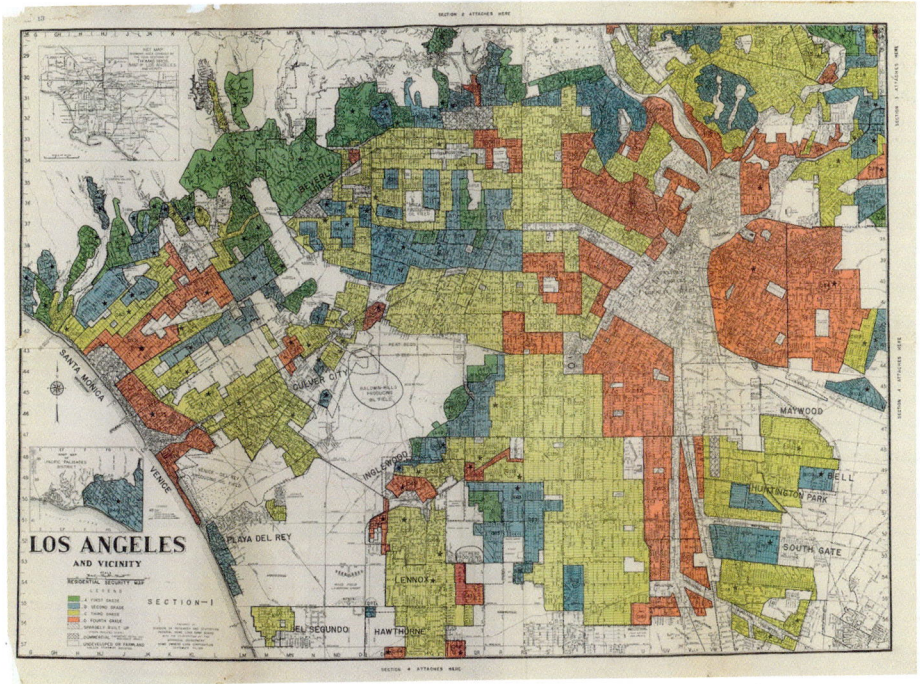

This 1930s map of Los Angeles was produced by the Home
Owner's Loan Corporation, or HOLC, a federal program
aimed at helping Americans recover from the Depression by
providing mortgage help. The program mapped 200 cities
based on the perceived risk of lending money—based in part
on demographics like race, ethnicity, and income. The resulting
color-coded maps are the origin of the term "redlining,"
and their impacts were both devastating and enduring.
National Archives

To do this, it mapped more than 200 US cities based on the perceived risk of loaning money in various areas, grading neighborhoods from A to D and outlining them on maps in corresponding colors, from green to red. Grades were based on the condition of the housing stock and on the area's demographics—the race, ethnicity, and income characteristics of residents. The result, as one study put it, was that "areas with predominantly US-born, white populations, and newer housing stock were often codified as the 'safest' places for banks to invest," and those places received grades of A and B. Places with older homes and more immigrants were graded C. And neighborhoods with "significant numbers of racial and ethnic minorities, foreign-born residents, families on relief, and having older housing were almost always viewed as 'hazardous' and given the lowest grade, D." Those were outlined in red, which led to the term "redlining."

The impacts of redlining were devastating—and they endure today. The policy "helped codify and expand practices of racial and class segregation," in the words of *Mapping Inequality: Redlining in New Deal America.* Ninety years later, nearly three-quarters of the redlined neighborhoods are still struggling financially, and nearly two-thirds are "majority minority," according to a 2018 study. Redlining's human legacy is vast: poverty, unemployment, health problems, decades of lost wealth creation and missed opportunities.

Environmental justice advocates have long sounded the alarm about the uneven distribution of pollution, with marginalized communities—most often non-White and immigrant—living in the shadows of oil refineries, chemical factories, power plants, and other sites that spew toxins into the air and water. These are often connected to redlining, either because their presence made an area cheaper, which brought people with no other option, which then led to redlining, or because redlined communities lacked the resources and political power to fight

off a polluter. But the policy also left less obvious ecological fin-
gerprints on many cities, effects that urban ecologists are now
eagerly bringing to light.

"There are just more people who have hardcore wildlife
training who are starting to look at cities as a place to do their
work," said Eric Wood, an ornithologist and urban ecologist
with California State University in Los Angeles and the Natural
History Museum of LA County. "If you'd told me as a PhD student, 'You, go study birds in LA,' I'd have said, 'No way, I'm going
to Borneo.'" Now, though, urban ecology feels like a discipline
brimming with the potential to make discoveries with real-
world impact. "Now people are saying, 'Heck yeah, I want to
work in the toughest neighborhoods.'"

Wood moved to LA in 2015 from the Cornell Lab of
Ornithology, eager to apply his field skills in a big city. "I'm a
birder and a natural history person, and for twenty-five years
I'd gone out and identified all the birds and plants and insects.
My bookcase is filled with field guides. When I got this job,
I was like, I'm gonna do all that stuff, but in the city—within-city
biodiversity." To measure biological diversity in a landscape, an
ecologist needs to capture the scope of environmental variabil-
ity there. In a natural setting, that might mean looking at hillsides that face north versus south, or areas with wetter or drier
soil, and so on. In LA, Wood soon found, the environmental vari-
ability was also based on neighborhoods' socioeconomic status.

"The differences we see in LA are in your face," Wood said.
"You can just go across the street from one of the most wealthy
places to a quite seedy part. The very roots of LA and of every

---

A great blue heron surveys the Los Angeles River. The river,
which has undergone a major restoration effort in recent
years, hosts many bird species as it winds through the city.
Herons fish in the river but nest in nearby parks. *Citizen of
the Planet/UIG/Alamy*

big city in the US are segregationist." In newer cities in the West, the residential segregation was not explicit, as it was in the South, but rather linked to designating neighborhoods as hazardous for banks to invest in. It's a legacy, said Wood, "of where the oil was in the city, where the African American people were working, and then there were pollutants in the ground, and then it was redlined. It's just been nonstop." It's a story about historical patterns of investment, he said, but also about systemic racism. Injustice is a barrier to opportunities of all sorts—and that turns out to be true for non-human species as well.

In LA today, the dominant ethnic group in the formerly redlined areas is Hispanic, whereas most of the areas that were graded A are largely White. In a study published in the fall of 2023, Wood and others surveyed birds across the sprawling metropolis and analyzed the findings against redlining maps. They found that predominantly White neighborhoods, which were often the ones "greenlined" on the HOLC maps, hosted a greater abundance of birds that generally live in forests, such as warblers, wrens, and bluebirds. In contrast, areas that today are predominantly Hispanic or Latino, which were most often redlined, have fewer of those forest birds and more synanthropic species, those often found in dense urban areas. (These included pigeons and sparrows but also crows and ravens, mourning doves, house finches, and even a type of hummingbird.) "Our results," the researchers wrote, "illuminate patterns of income inequality, both past and present, that carry over to influence urban biodiversity."

As examples, Wood compared Beverly Hills, where the average home price is more than $3.6 million, according to Zillow, with Boyle Heights, a largely Hispanic neighborhood where the average home price is $628,000; it shows up as a large red blob on the HOLC map and has far fewer trees and green spaces. "You get loads of these birds that require insects for their life history,

and they go to a place like Beverly Hills because there are trees and flowers," he said. The differing landscapes clearly matter to the birds. But while it might not immediately matter to people whether they share their neighborhoods with a common raven or a yellow-rumped warbler, there's a larger point: Places like Beverly Hills and Boyle Heights are only a dozen miles from each other, but worlds apart. The birds are "an indicator of these broader conditions that are effectively bad for people."

Another study, led by Dexter Locke, a researcher with the USDA Forest Service, examined the intersection between redlining and tree canopy across American cities. The study looked at thirty-seven US metro areas that were redlined and found that, on average, areas graded D have only 23 percent tree canopy today—even though redlining as a policy officially ended in 1968, with the passage of the Fair Housing Act. Areas graded A, by contrast, have 43 percent canopy. (In many D areas, there isn't even space to plant trees, whereas many A areas are characterized by single-family homes with yards.) Tree canopy is important both as vital wildlife habitat and as a means of reducing the urban heat island effect, something essential for urban dwellers of many species. (As I wrote this paragraph, Phoenix had just experienced its nineteenth straight day of temperatures over 110 degrees, and the mercury was 100 degrees or higher in cities around the world, from Rome to Kyoto.) Trees are crucial to preventing heat-related deaths— and they are becoming more so by the day.

At a meeting of urban wildlife researchers I attended in the summer of 2023, a diagram from Schell's 2020 paper— illustrating the links between systemic racism, landscape effects such as urban heat islands, and impacts to biodiversity—made it into so many PowerPoint presentations that its recurrence became a running joke. The paper sparked a flurry of trail-blazing studies. Chloé Schmidt was a graduate student at the

University of Manitoba in 2020, and for her PhD research she had been assembling a database of genetic information about biodiversity—"all the big genetic data I can find, to address broad-scale questions," she said. When she read Schell's paper, she realized she had the data to test his ideas. "Redlining was so consistently practiced for so long in the US, we thought we could find a signal," she said. She set out to see whether redlining left an evolutionary mark on urban wildlife.

Using genetic information from nearly 7,700 individual animals belonging to 39 different species of vertebrates, Schmidt showed that across 268 urban locations in the US, the wildlife in neighborhoods with greater proportions of White residents had higher levels of genetic diversity—essential for wildlife populations to weather a catastrophe like a pandemic or a wildfire—and more evidence of connected populations of animals. The finding revealed a blunt truth: Just like a wall or fence or a highway, systemic racism creates a barrier to wildlife movement.

Schmidt, who went on to work as a senior scientist at the German Centre for Integrative Biodiversity Research, is of mixed race, and when she was growing up in New Jersey, her parents sometimes mentioned that the original deed to their house "said Black people couldn't live there." Even as a kid, she said, she was always aware of "the differences in towns and neighborhoods when you're driving around." But learning all about redlining, and seeing its impact on wildlife, too, was "a mind-blowing thing." "The whole process changes your view of the world, honestly," she said of the research. "It was like, 'Oh, god, how bad must this have been to still find a signal even when redlining was stopped in the '60s?'"

— – – —

In Durham, Katti, the ecologist from NC State, was also studying the complex socioecological relationships among people and wildlife in cities. He and Jin Bai, his PhD student at the time,

who are both birders, wanted to see whether they could detect signals of social inequality in the patterns of bird distribution across Durham—similar to the work that Wood was doing in Los Angeles.

I had first met Katti more than a decade earlier, when he was based in Fresno. Back then, he was studying the luxury effect by looking at the way a household's income influenced its outdoor water use and thus its landscaping, and how that in turn influenced the diversity of birds in the area. I wrote about his work, part of a national effort to better understand how urban ecosystems function.

At the time, wealthier households in Fresno were using more water in their yards, and those landscapes also hosted more biodiversity. Fresno residents paid a flat fee for their water regardless of how much they used, so the water use wasn't linked to the cost of the water. It was more that people with more money had a preference for lusher landscapes—and perhaps also the resources, like time, money, and home ownership, required to either garden or think about gardening and hire someone else to do it. But the city of Fresno was about to start charging residents for their actual water use, in the hope that metered pricing might reduce water consumption in an era of dwindling water supplies; nearly three-quarters of residential water use in Fresno went toward irrigating yards and gardens. Katti hypothesized that this change in policy would cause a decline in local bird diversity, since watering less would decrease the diversity, size, and number of plants in yards.

I was intrigued at the time by the potential unintended consequences of both urban policy and daily actions like watering a yard, and by the notion that we could uncover previously unseen relationships simply by measuring and comparing things as basic as water usage and shrub density. (In reporting that same story, I learned that researchers measuring greenhouse

gas emissions in Boston had discovered a "weekend effect," a decline in the amount of carbon dioxide emitted in the city on Saturdays and Sundays. One scientist had poetically called it "a pulsing type of urban metabolism.") This simple relationship between birds and human water consumption is just one tiny example of the interconnectedness of humans and nature, and a reminder that cities aren't exempt.

One summer morning, I joined Katti and Bai in a lush park in Durham. As the smoke from catastrophic wildfires in eastern Canada that year turned the air a soupy gray, we took turns having coughing fits as we listened to the calls of cardinals and robins, chimney swifts and fish crows. For his PhD dissertation, Bai was studying the impacts of redlining on bird abundance and diversity—what kind of birds and how many—in parks in areas of the city that had been zoned A and B versus those that were zoned C and D.

We stood in the smoky haze that summer morning in an otherwise lovely park in an AB zone, with a creek running through it and a wall of towering trees along one side. Katti pulled up a Zillow map on his phone. "These are half a million to $1.6 million on this side here, and across the stream they're all over a million dollars," he announced, pointing toward some stately houses opposite the park. We watched as a White family loaded kids into an SUV, headed for some summer activity or another.

Bai , who is now a research associate at the Nicholas Institute for Energy, Environment & Sustainability at Duke University, had recently collected bird data in Durham parks, documenting

---

A century ago, much of North Carolina was farmland. But the shift to a more urban lifestyle brought trees back to the landscape. Today, the Research Triangle area—encompassing Durham (opposite), Raleigh, and Chapel Hill—is more forested than many urban areas across the US. Differences across cities in terms of available green space and other natural resources influence wildlife movement. *Kevin Pugh Media*

every bird he observed within forty meters during a five-minute period. Because other studies, like Wood's, have found that redlined areas have lower biodiversity, and because it's often true that more trees leads to more birds, Bai had expected to see these same patterns. So far, though, he hadn't found what he'd expected. He found, generally, the same number of birds across all the parks, and largely the same types—with just a few exceptions. The findings showed the danger in making universal assumptions; all cities are not created equal, so it made sense that they wouldn't all behave the same way, even in the face of a common policy like redlining. As Schell put it, "Seattle's not Tacoma. Oakland is not San Francisco. San Francisco is not Los Angeles, right?" Cities are fundamentally different from one another.

Bai's study specifically looked at parks, whereas Wood's had looked far more widely throughout neighborhoods. Bai thought that the lack of difference he'd documented in the bird diversity showed how essential parks are in cities. "All of these parks are super important in urban environments," he said. "They may serve as a refuge for birds." Birds were attracted to the parks no matter which zone because they contained vital resources; all the parks served this same role.

Other recent research backs this up. Amy Vasquez, a graduate student who worked in Wood's lab, looked at bird diversity in parks across Los Angeles and found that they served as refuges for birds in areas that generally had fewer parks. Even in areas that had fewer forest birds and "otherwise inhospitable urban conditions"—places like Boyle Heights—the diversity of birds in the parks themselves was similar to parks in greener, Beverly Hills–type areas. If the habitat is there, birds will find it. Unlike with mammals that need on-the-ground corridors to move through, it can be fairly easy to create connectivity and habitat for flying species in urban areas. It just takes the will to do it.

I wanted to see what a park in the formerly CD zoned areas of Durham looked like, so we headed to another park just a few minutes' drive away. It still seemed lush, especially from my arid Colorado perspective, with large open areas and a wall of forest along one side. This park had more concrete and infrastructure—basketball courts, a baseball diamond, a picnic shelter. But it still seemed like a nice place to be a bird. Zillow told the human story: Here, houses around the park cost between $100,000 and $400,000, well below the home values in the first park we'd visited.

Looking around, though, it was also clear that gentrification was coming. Interspersed among houses and yards in need of repair and tending, there were new homes going up, sleek boxy structures that looked like townhouses, but freestanding. Gentrification casts its own shadow over humans and nature. It brings a new kind of discrimination to neighborhoods, pricing out communities that have long lived or run businesses there but can no longer afford the rent. For wildlife, it can dramatically alter the available habitat, for better or worse. New property owners may ultimately increase the diversity of plants and flowers, contributing to a luxury effect. But not always, especially in the short term. They may want a bigger house, and will cut down trees to build it. Newly valuable lots might be subdivided, leading to loss of green space. Vacant lots are unlikely to remain that way as real estate values rise.

Katti and Bai pondered the intersecting forces. A century ago, much of central North Carolina was farmland, with much less tree cover than today. "What happened here is the shift from agricultural to urban means people became more concentrated, the pressure on the land actually went down, and trees came back up," Katti said. (Unlike in the West, green cover can quickly grow back in lush Eastern landscapes, which may also help explain Bai's findings. Wood called it "feral landscaping.") The Research

Triangle—the Raleigh–Durham–Chapel Hill area—is relatively heavily forested compared to most metro areas; Raleigh is known as the City of Oaks. Today, though, the opposite phenomenon may be happening, with people cutting down forest patches to build more housing. "It is very actively dynamic," said Katti, "so it's hard to think about the luxury effect or these kinds of correlations as being static."

We visited one more park in a formerly redlined area, and this one was far less bucolic. Located in a more industrial part of town, it was bordered on one side by a noisy road where trucks rumbled by frequently. On another side, a thin row of trees atop a small slope separated the park from an asphalt lot that housed a landscaping company. Flocks of sparrows and starlings sat atop a dumpster there. This park was the exception in Bai's study: It hosted higher concentrations of synanthropic, non-native birds, more akin to what Wood had found in LA's formerly redlined areas. Still, it also contained patches of forest, and Bai had seen some more exciting and hard-to-spot bird species here, including a yellow-billed cuckoo, an American redstart, and a magnolia warbler.

The birds were there, in fact—but interestingly, the birders weren't. Not only does the composition of wildlife differ between neighborhoods, but so does the incidence of people looking for wildlife.

— – – —

American cities are relatively young, and loss of genetic diversity from loss of habitat occurs over many generations of animals, "with slow but increasing rates of genetic erosion through

North Carolina State University ecology PhD student Jin Bai saw a magnolia warbler, as well as other hard-to-spot birds, in a park in a relatively industrial—and formerly redlined—section of Durham. Parks can serve as essential refuges for wildlife in cities. *Alan Murphy/BIA/ Minden Pictures*

time," as Chloé Schmidt wrote in her study. It's similar to the idea of extinction debt—the idea that habitat loss now will lead to extinction in the future, though there may be a time delay between the cause and the effect. Since the Industrial Revolution, wildlife across the planet has lost an average of 6 percent of its genetic diversity. Redlining is a more recent phenomenon (though residential segregation in other forms predates it, of course); its evolutionary effects are percolating through urban wildlife populations but are not yet set in stone. "There is still time to make positive change with environmental interventions that promote gene flow from more genetically diverse populations across the urban racial mosaic," Schmidt wrote.

One way to spur that change is to recognize and remedy the related problem of inequity in wildlife observation. Diego Ellis Soto, a postdoctoral researcher at Berkeley, looked at the amount of available biodiversity records and discovered that across the country, historically redlined neighborhoods had the fewest records available. Those records are "the first building block to make any biodiversity conservation decision," Ellis Soto said. They are the data that underpins a host of policies and priorities, from conservation funding to infrastructure planning. Ellis Soto found that neighborhoods that had been graded D had 74 percent fewer bird observations than those graded A. According to the study, published in 2023, "formerly redlined areas are significantly undersampled compared with non-redlined areas, and ... there is overall less certainty about the number of bird species that are known to occupy these areas."

That discrepancy could significantly impact conservation agendas. "How can we protect what we don't have information for?" Ellis Soto said.

Ellis Soto, who is from Uruguay, said that when he first arrived in New Haven for a PhD program at Yale, he was shocked to see how segregated the city was. "People live in a different

New Haven and a different reality from street to street." The inequalities in data collection in those sorts of different realities are also growing: In the past two decades, the gap in available records grew by 36 percent. If you want to make planning decisions that can make cities more wildlife-friendly, you need to know where the wildlife actually is. You need to know which birds are using which parts of urban landscapes.

Data from national bird surveys like Audubon's Christmas Bird Count and the popular birding app eBird are "skewed spatially in representation of higher-income neighborhoods," Katti said. While eBird does get a lot of data from urban sites, "it's very patchy and unequal sampling."

When Katti lived in Fresno, he started a local bird count, a citizen science effort to map birds around the city. He developed a method that was distinct from the Christmas Bird Count. He later moved to Phoenix, and when he started a bird count there, he went looking for data from the oldest Audubon counts to use as a comparison. But he discovered that because volunteers chose their own locations, the data didn't come from the same spots year after year. Instead, the counts tended to cluster in circles around the city's edge, and as the city grew, the circles moved outward—not unlike the White suburbs that sprung up after redlining, as more privileged people moved out of urban centers. "It's no use for the city birds, because you don't have continuity of the same location," Katti said.

The methodology for Audubon's annual Christmas Bird Count varies by location, but in most places, coordinators divide a count area into blocks and let volunteers pick a location anywhere within that block. "It's all up to the volunteer where they want to visit," said Bai, "and most likely, they want to visit somewhere more natural like a preserve, or somewhere away from people." If you're going out for a morning of birding, you're unlikely to go to the park by the train tracks with the trucks rumbling by.

Unless you're Bai, that is. He pulled up a map on his phone from a different bird mapping tool, a state atlas of breeding birds in North Carolina. It showed every location where someone had submitted a birding checklist. You could see that in Durham there were way more checklists in the AB zones than the CD ones. And some of those CD lists had been submitted by Bai himself.

"The point we're trying to make," said Katti, "is that to use it for any kind of planning decisions, the dataset is not reliable." Katti's bird count, which he now runs in the Research Triangle, instead divides an urban area into one-square-kilometer grids and then randomly draws one point in each grid—and that's where the volunteers go. (If that random point ends up being on a rooftop or someplace inaccessible, then they move it to the nearest accessible point.) While the Audubon and eBird counts are focused on finding as many birds as possible—"Those counts are motivated more by going where the most birds are," Katti said—Katti is interested in how birds populate more-urban environments: "Our motivation is to see what birds are where the people are," he said.

Data collected by citizen scientists and recorded on eBird is widely used in scientific research and conservation action. On a popular open-access international database called the Global Biodiversity Information Facility that provides information about species all over the planet, eBird observations make up nearly half of all the data. (Other information comes from sources as disparate as eighteenth-century museum collections and modern DNA barcoding. Ellis Soto's research showed that museum collections are similarly biased.) Information from

---

In Durham, North Carolina, Madhusudan Katti watches a blue jay harass an American kestrel. Katti and his student Jin Bai studied the complex socio-ecological relationships among people and wildlife in cities. *Cornell Watson/The New York Times/Redux Pictures*

eBird is used to make conservation decisions about everything from habitat restoration to captive breeding to allowing or prohibiting development. Gaps in bird observations based on socio-economic factors could have huge implications.

Models that conservation organizations use to identify where their work could "give the biggest bang for the buck," Ellis Soto said, may well be generating the wrong answers if they aren't based on accurate data. "If we don't know where the birds are, it's going to spit out different pixels." Ellis Soto also studied data on insect species, including monarch butterflies and mosquitos, using records from California and Texas, and found the same lack of data-gathering in marginalized communities. The data disparity is both a conservation and a social justice issue, he said. "If we sample mosquitos only where rich people live," that influences our understanding of where mosquitos are now, where they may go under climate change, and where disease might be distributed in poorer, often BIPOC, communities. "The more and more I think about this whole thing," he said, "having this biodiversity info will increasingly become important for political reasons."

Ellis Soto was volunteering in New Haven schools as part of Pathways to Science, a program that connects underrepresented students with STEM opportunities. He was working with kids to make hip-hop and bachata music from bird songs— "bluejay bachata!"—as a way to connect them to the wildlife sharing their space. Bai was working with the local Audubon chapter to try to diversify its membership—a big effort underway among birders across the country. Katti hoped all of this work would push city governments to start paying attention to the way structural injustices impact the natural world.

Schell and Wilkinson, meanwhile, hoped they could shed light on how the systemic injustices we perpetrate on our own species trickle down to other animals that share our space,

using coyotes as a model. "If we want to understand the mechanisms that led us to a very uneven landscape of where the biodiversity sits, then we need to necessarily understand the lives of these animals," Schell said. "That's our next step. What are the lives of these individual animals? How are they coming into contact with these contested spaces? What does it mean to start drawing comparisons between these organisms, seeing them as what we would call bioindicators or ecosystem sentinels?" A graduate student in Schell's lab, Tal Caspi, collected hundreds of fecal samples from coyotes across San Francisco. In the lab, they were analyzing the glucocorticoids—stress hormones—as well as other hormones, and connecting that metabolic information to the animals' diets and home ranges to see how their food and habitat were influencing their stress levels. "There are a lot of environmental disturbances we have deposited onto the landscape—noise, light, air, pollution, heat, plastics," Schell said. "All of that is having a consequence, and we don't fully understand it."

What these ecologists do understand, though, is that our own well-being and that of wildlife are inextricably linked. The close confines of cities amplify this truth. "I think we get so anthropocentric in our world view that we completely disregard all of the damages done to every other organism around us," Schell said. "Why on Earth would they somehow be immune?" We are all connected. Why is it so easy to forget this obvious fact?

Even at the edges of cities, where human society bumps up against wilder realms—a region often referred to as the wildland-urban interface, or WUI—the two worlds bleed together and interact in unexpected ways. There is no impermeable line between "human" and "natural"—not in cities, not around national parks, not along roads. And not in the flows of water hidden underground—which I would learn firsthand at a bird sanctuary in the Everglades.

Hidden Connections:
In the Everglades,
Water Is Everything

Shawn Clem had just started to tell a story when her right front tire slammed into a rut. She tried to reverse out, but the pick-up's wheels just spun. She tried going forward again. No dice. She cautiously opened her door and stepped out to take a look.

We were atop a small culvert on an extremely narrow section of a narrow dirt road across a swamp—which was filled with several feet of water and a whole lot of alligators. It was definitely not where you wanted to slide off the road. I rolled down my window and leaned out to peer at the rut directly beneath me, just in time to see an alligator waddle into the water, so close I could almost have reached out to pet it. Standing next to the truck in the sweltering July heat of southwestern Florida—93 degrees and about 85 percent humidity—Clem,

---

Previous Spread: Conservationist Marjory Stoneman Douglas famously referred to Florida's Everglades as a "river of grass." This unique wetland ecosystem once covered the entire southern portion of the state, from Orlando to the Keys. But since White people arrived, most of its history has involved attempts to destroy it. *Oliver Rogers*

the director of conservation for Audubon's Corkscrew Swamp Sanctuary, pointed out another alligator just ahead of us under a tree. There might be a nest there, she said casually, getting back in the truck. Sure, nothing to worry about.

I was glad Clem was at the wheel. The rut turned out to be fairly small, and there was really nothing to do but plow forward. She switched the truck into all-wheel drive and tried again. After a bit of rocking and rolling, we were moving. A few hundred yards ahead, a wooden tower straddled the dirt track. We parked the truck and climbed the stairs. From atop the tower, the marshland of the Corkscrew Swamp Sanctuary spread out in every direction. Here, on the western side of the Everglades, we could see the "river of grass," as pioneering Everglades conservationist Marjory Stoneman Douglas described it in the 1940s, stretching beneath us. It was a sea of sparkling greens of every shade, grasses and sedges and taller shrubs rising above the shimmering marsh, with small patches of oak forest—little elevated islands called hammocks—scattered throughout. In other places, the land gave way to ponds whose water reflected the puffs of white cloud gathering in the otherwise pale-blue sky. It looked idyllic, nature functioning as it always had. But looks can be deceiving.

The Audubon preserve spans about 13,500 acres just east of Naples. It contains some of the oldest and biggest bald cypress trees anywhere in the world, in the largest remaining intact bald cypress forest, 700 acres of old-growth trees, some more than half a millennium old. The sanctuary is home to an assortment of amazing species, many of which are listed as federally or locally threatened or endangered: American alligator, white ibis, roseate spoonbill, Big Cypress fox squirrel, gopher tortoise, Florida panther. Among its most celebrated inhabitants are wood storks, an endangered species of wading bird. These elegant dinosaurs stand four feet tall and have bumpy bald

necks and heads, long beaks, and white plumage with dramatic black-tipped wings. They were hunted nearly to extinction in the early 1900s, when their feathers, along with those of egrets and herons, were prized for women's hats.

In 1912, the National Audubon Society hired a warden named Rhett Green to live at Corkscrew Swamp and protect its nesting colonies from plume hunters' slaughter. The unflappable Green threatened to "shoot on sight" any plume hunters in the vicinity, and his doggedness, along with a national campaign against the inhumane practice, ultimately succeeded in saving what was left of the species. The bird population eventually stabilized, but it was a shred of its former self: The number of wood stork nests had plummeted from 100,000 to perhaps 10,000.

Then Corkscrew's wood storks faced another crisis. In the 1940s, Florida picked up the pace of decimating and draining the Everglades, which had been happening in fits and starts ever since Florida became the twenty-seventh state in the Union in 1945. By the 1950s and '60s, southwest Florida, where Corkscrew is located, was in a development frenzy that obliterated much of the stork's wetland habitat. In the early 1950s, timber companies arrived to log the giant old-growth cypress, much of which was sent to Europe to rebuild its cities in the wake of war. Loggers "were taking down all the old-growth cypress literally up to our back door," Clem said of Corkscrew Swamp. A heated grassroots effort to save the cypress resulted in a huge success: A timber company sold a chunk of the swamp to Audubon for $100,000, raised from more than 200 individual donors, organizations, and businesses. A few years later, a second company agreed to sell an even bigger parcel for just

A wood stork feeds in the shallows near the Loxahatchee River in Jupiter, Florida. North America's only species of stork, these birds are tactile feeders—they use their feet to find fish, mollusks, and other aquatic food. *Sydney Walsh*

$25,000. Today, along a two-mile boardwalk trail through the Audubon preserve, signs celebrate "landmark" cypress trees—including one called Asteenahoofa, the Seminole name for big cypress, which "contains an estimated 32,000 board feet of merchandisable lumber," according to its placard. "It would have been prized by loggers whose axes were halted by Audubon less than a half mile south of this tree." Ecological disaster had been averted at Corkscrew, twice.

But as Naples and the surrounding areas of southwest Florida grew, the wood stork's habitat continued to shrink. Now it's clear that the sanctuary and the creatures that live there are at risk again, this time from more complex threats. In the early and mid-twentieth century, conservation problems, though devastating, seemed fairly straightforward—poachers are killing all the birds, loggers are chopping down all the old trees—and they often had clear, direct solutions. (Which is not to say, of course, that those solutions were easily attainable. But at least there was direct cause and effect.) Protecting the birds from poachers and the trees from loggers "were physical things we could do," said Clem. Despite having triumphantly protected the birds and the trees decades ago, though, today Audubon is finding that neither of those actions is enough.

Wood storks are vanishing from the sanctuary. In some years, they are absent entirely. "Last year, we did get nesting," Clem told me in the fall of 2023. "We had about fifteen nests. But a really good nesting year for us right now is one hundred nests, and we haven't seen that in a while. When we get nesting now it's a dozen or two, and more often than not there's none."

The reason why has to do with hydrology—how water moves through a system. And it illustrates the way in which connectivity is about more than just corridors between protected areas. It's about functioning systems—in this case, one that involves water moving in often invisible ways. "We can protect what's

inside our boundaries to a certain extent," Clem said of the sanctuary's mosaic of marshes, oak hammocks, pine uplands, and cypress groves. "But the factors affecting us are *outside* our boundaries. That's the challenge we're facing now."

It's an increasingly common predicament: How do you safeguard the natural world when the forces affecting it are geographically separate and sometimes many steps removed from your sphere of influence? It is the climate change quandary on a local, visceral scale. It shows the true breadth of connectivity, the complexity of interconnected systems in the natural world, and the additional complexity involved when you add human development to the mix.

"We're losing the connections of the system," Clem said. "Even though we have wetlands like the sanctuary remaining, we are losing water." Despite Corkscrew's history of success, trouble has arrived once more on the sanctuary's doorstep.

— – – —

Florida's Everglades was once a vast natural wetland ecosystem that covered the entire southern portion of the state, from Orlando to the Keys. "Nothing anywhere else is like them," wrote Marjory Stoneman Douglas in her book *The Everglades: River of Grass*, first published in 1947. "Their vast glittering openness, wider than the enormous visible round of the horizon, the racing free saltness and sweetness of their massive winds.... The miracle of the light pours over the green and brown expanse of saw grass and of water, shining and slow-moving below, the grass and water that is the meaning and the central fact of the Everglades of Florida. It is a river of grass."

Next Spread: A wood stork nesting colony in Florida's Caloosahatchee River. Wood storks were listed as endangered in 1984. Since then, intensive recovery efforts have helped the population to double—but the birds require landscape-wide conservation planning that includes ensuring just the right seasonal water levels. *Jacob Zetzer/Audubon Florida*

The Kissimmee River, the feeder for that river of grass, once snaked and meandered across a wide floodplain before emptying into Lake Okeechobee and then spilling into the Everglades, percolating through the seemingly endless wetlands, and finally meeting the sea. But virtually since White settlers first colonized the area, the history of the Everglades has involved misunderstanding and maligning the whole system as a miserable swamp whose only value lay in its vanquishment, and attempting to harness it for human gain. As Douglas put it, for "four hundred years the region now called 'The Everglades' was described as a series of vast, miasmic swamps, poisonous lagoons, huge dismal marshes without outlet, a rotting, shallow, inland sea, or labyrinths of dark trees hung and looped about with snakes and dropping mosses, malignant with tropical fevers and malarias, evil to the white man."

Ultimately, Floridians succeeded in their bid to overcome nature—if you can call something that wiped out one of the world's most unusual and exquisite ecosystems a "success." For decades they dredged and built canals and straightened the river into concrete channels and diverted water with dikes and weirs, obliterating the lush and teeming wetlands, the sawgrass, the mangroves, the oak hammocks. Human development destroyed half the land in the Everglades and 70 percent of its water flows. But then a monumental national effort to restore the Everglades officially passed in Congress and was signed by President Bill Clinton in 2000, as law and policy finally caught up with what small bands of conservationists and environmental activists and the occasional politician had been shouting about for more than half a century. It's the world's biggest ecological recovery effort, commandeering the might of the US Army Corps of Engineers, which built much of the infrastructure that destroyed the system, to revitalize it—by restoring the river's natural bends, elevating portions of a highway to let

water flow beneath the road, building reservoirs to hold water and release it slowly, and many other interventions.

Each year since 1986, an alliance of nearly sixty conservation groups known as the Everglades Coalition has hosted a meeting on the status of the ecosystem and the progress toward their restoration goals. The conference's annual titles read like a history of the movement's agonizingly slow but tangible headway. The thirteenth meeting, in 1998, was "Strategies for Success." The following year, responding to an apparent stall in momentum, the title was "Time to Get the Word Out." In 2004, in another attempt to accelerate the pace, it was "Providing the Leadership, Renewing the Partnership." Two years later, it was a seemingly dejected "Are We Making Progress?" Other titles over the years reflect an emphasis on shifting priorities or attempts to gain traction—protecting coastal communities, building connections, investing in climate resilience. But in 2023, the conference title took a decidedly upbeat turn: "A Watershed Moment for America's Everglades." I decided to attend.

I arrived in Fort Lauderdale late on a Thursday night, driving my rental car down the river of asphalt known, apparently without irony, as the Sawgrass Expressway. I passed the exit for the Sawgrass Mills Mall, wedged between a Walmart and a Sam's Club, and arrived at the slightly run-down conference hotel that turned out to be just over a mile from Marjory Stoneman Douglas High School—named for the grand dame of Everglades conservation but now known, heartbreakingly, as the site of yet another mass shooting of children. The Marriott parking lot was full, but I found a spot along the crumbling pavement in a far corner, where the sound of my car doors startled a raccoon that was sniffing around beneath a palm tree. I checked in, ate some pretzels for dinner, and crawled into bed feeling less than optimistic about Florida's, or anyone's, future.

In the morning, I discovered that the rest of the conference attendees were in a different mood. These folks, who hailed from local, state, and national conservation organizations, state and federal agencies, tribes, and universities, seemed quite pleased to be mingling in a moldy hotel ballroom, thrilled to see one another, and brimming with energy and optimism about their work. I soon learned one reason why. In one of the first acts of his second term earlier that month, Governor Ron DeSantis had committed an unprecedented $3.5 billion for Everglades restoration.

Conservationists I spoke to saw it as a calculated move to support a wildly popular issue in advance of a presidential campaign. In DeSantis's three terms in Congress, the League of Conservation Voters had put his environmental voting record at 2 percent. But these conference attendees had worked too hard for too many years to look a gift horse in the mouth— even a politically calculating one. There was serious work to do in the Everglades, and now they had the money, more than they'd ever dreamed of, and the political backing to do it. In Everglades National Park, which occupies more than 2,400 square miles of land and water at the southwestern tip of the state, south of Corkscrew Swamp Sanctuary, things were looking up, thanks to increased delivery of water from the upper parts of the Everglades system. Water depths were increasing, and so-called dry-downs, when the peat dries out and the risk of fires goes way up, had recently been shorter and less severe. Sawgrass was growing back, reflecting an important shift in

---

Top: In the 1960s, as part of a gargantuan project to drain the Everglades for farming and development, the US Army Corps of Engineers channelized the Kissimmee River, which proved to be an ecological catastrophe. *Bottom:* Between 1999 and 2021, forty square miles of the river's floodplain were restored through a federal-state partnership. *Both photos courtesy of the South Florida Water Management District*

vegetation. Alligators were nesting in a wider range of areas—
though park officials cautioned that it was too early to attribute
this yet to the changes in water levels. And wading birds were
also moving back to their historical nesting areas. "All great
news from the park!" was how Melodie Naja, the director of
the National Park Service's South Florida Natural Resources
Center, summed up the news during a "status report" panel.

Of course, there were still challenges—big ones. Conservation
gains were proceeding alongside large amounts of new development. The state's population had exploded, from about 1.9 million back in 1940 to more than 22.2 million in 2022. On average,
more than 1,100 additional people were moving to Florida every
day. That's more than 400,000 new residents, or nearly another
Miami's worth, each year. Where were all those people going
to live? And while there was real momentum around restoring
natural water flows through the Everglades and constructing a
wildlife corridor through the state, there was still, simultane-
ously, a huge amount of development proceeding in important
wildlife habitat, and a seeming lack of any means to stop it.

Meanwhile, climate change is adding an extra layer of challenges and chaos, as sea level rise spurs species to move in
search of new homes, drought raises the risk of wildfire in some
places, floods destroy nesting sites, exotic species and diseases
wreak havoc, and salt-tolerant species move inland, changing
the mix of vegetation and forcing still more species to move
in search of new habitat. Encroachment, climate change, sea

Audubon Society staff on high ground at the Corkscrew
Swamp Sanctuary in Naples, Florida, in March 2024. During
the dry season, from October to May, water levels slowly fall,
and fish become concentrated in dwindling pools. This makes
prime feeding conditions for wood storks and other tactile
feeders, like ibises and roseate spoonbills. When water levels
are high in the rainy summer season, visual feeders like herons
and egrets thrive. *Sydney Walsh/National Audubon Society*

level, invasive species: "Any of these could occupy all your time on its own," Shannon Estenoz said during the meeting. A conservationist from Key West who had spent most of her career on Everglades restoration, Estenoz was appointed as the US Department of the Interior's Assistant Secretary for Fish and Wildlife and Parks under President Biden, another sliver of hope for the Everglades.

I watched a presentation by biologist Craig van der Heiden, who spoke of how difficult it was to manage wildlife in such a rapidly changing system. As an example, he showed a photo of two very skinny deer, their hip bones sticking out, standing in water up to their knees. White-tailed deer, along with dozens of other species, have found themselves unable to adapt to the disappearance of tree islands in the Everglades.

The islands, just two or three feet above the sloughs (shallow, marshy lakes—strangely pronounced *slews*), are critical habitat for biodiversity. In the natural system, they might disappear underwater during a flood but would quickly reemerge. But in an altered, dammed system, these islands can disappear underwater for weeks or months, leaving wildlife with nowhere to go. Softshell turtles and snakes used to be much more common on tree islands, but now their eggs can get washed away by flooding at the wrong time of year. Gumbo limbo trees—large shade trees with red bark that are important canopy species on tree islands—are now almost gone from the central Everglades because they can't survive long periods of inundation. "If we don't have corridors, ways for animals and vegetation to move, they will become locally extinct," van der Heiden said.

The Miccosukee Tribe—for whom van der Heiden works— once lived in northern Florida, on Lake Miccosukee, but about 100 members fled to the Everglades when Andrew Jackson invaded their territory in 1818. They survived by living in small camps on the hammocks. Today, the more than 600 tribal

members are descended from those refugees, and they are dedicated to helping restore the Everglades. "The Everglades is our mother," William J. Osceola, the tribal secretary, said at the Coalition meeting. "The most important thing alongside family." The Miccosukee are committed to restoring the Western Everglades—where Corkscrew is located—which was home to the earliest Everglades restoration project but still hadn't received the same attention as other parts of the system.

Everyone hoped that was poised to change. It would need to, for Corkscrew and its wood storks and the larger ecosystem to thrive.

— — — —

The defining feature at Corkscrew, as in the Everglades more broadly, is water. Summer in Florida is the rainy season, which is when the swamp, like a giant bowl, fills up. Then, during the dry months, from October to May, the water levels slowly fall, with the highest-elevation areas—we're talking differences of a foot or two in elevation—drying up first. Populations of fish and other aquatic life become concentrated in smaller and smaller pools, and birds and other wildlife flock to those pools for food. When the water levels are high in the summer, herons and egrets can flourish—in July, I saw them wading, dining, and soaring—because they are visual feeders. They can see a fish beneath the surface and grab it. But birds like roseate spoonbills, ibises, and wood storks are tactile feeders—they catch fish by touch. When water levels are high, these birds expend more energy finding a fish than they would gain from eating it. They also can't feed effectively when they have to put their

Next spread: The Kissimmee River once snaked across a wide floodplain before emptying into Lake Okeechobee and then spilling into the Everglades, percolating through the seemingly endless wetlands, and finally meeting the sea. The Kissimmee River Restoration is the largest successful river restoration project in history. *Carlton Ward Jr.*

heads underwater. So in the summer, those birds live farther north in Florida and up into southeastern Georgia. "They won't really move in until the water levels start to go down and the prey are more concentrated," Clem explained.

Around December, conditions are right for wading birds to nest. "They're relying on the long, slow recession of water to create high-density pools of fish," said Clem. "And they need that food source to last throughout the whole nesting season, which goes until May or June. I think of those pools of fish like little stoplights that blink on and off." The "high" elevation pools will dry up early in the season, followed by those in the middle elevation, and finally the deepest parts of the swamp. "You need all those different types of habitat. The only way that's possible is if you have a system that's very connected, so the water can flow through those different habitats."

Clem calls wood storks "the Goldilocks of wading birds"—if the conditions are suitable for them, "you're helping the whole suite of wading birds," she said. When she joined the sanctuary staff back in 2006, the wood storks were entirely gone in some years. But back then the sanctuary didn't have a significant scientific research program, and Clem's work was largely focused on the Big Cypress Swamp, a federal protected area to the southeast. "We knew wood storks were declining, we knew it had to do with development—which it does, because they're declining across all of south Florida." Then came 2018, "a banner year for wading bird nesting in the Everglades," but with virtually no wood stork nests at Corkscrew. As a side project, whenever she had a few free minutes, Clem began looking at data related to the wood storks, including changes in water levels.

By 2020, it was clear something was going on. Clem, a problem-solver by nature, dove in. She looked at sixty years of data on water levels and discovered something shocking. There was far less water in the sanctuary than there used to be.

Starting around 2000, the dry season began drying out more quickly, while the so-called short hydroperiod wetlands, which filled with water for only one month of the year, were disappearing altogether. Climate change wasn't the culprit; this change was far more abrupt. But it was especially concerning because climate change will also have a drying effect here.

Recent modeling looked at how rising temperatures will impact the sanctuary's water, particularly through a process called evapotranspiration. Evapotranspiration is really two processes, which combine to move water from land to atmosphere. There's evaporation, when liquid water in surface pools or in the soil turns to vapor, and transpiration, when plants release water vapor through their leaves. Higher temperatures lead to more of both these things, drying out land, water sources, and plant tissue more quickly. As temperatures rise, the wet period across the whole sanctuary could decrease by about six weeks. "Habitats that are only wet for maybe ten weeks out of the year—I mean, that's only going to be wet then for four weeks, so it's a big change," Clem said. If the high-density pools of fish dry out sooner, wood storks lose their food source just when they need it most, when they are trying to raise their young and preparing to migrate north.

Water levels at Corkscrew Swamp go up and down by four to five feet throughout the year. Clem said she and her husband joked about how much wetter or drier it seemed each season than it did the year before, "because every year kind of resets your clock." But the system can also look different from one year to the next. "Our hydrology from year to year varies so much that it can mask big changes. We'll have a couple wet years and a couple dry years, it's a really dramatic system." Those fluctuations had been enough to mask the two-decade drying trend. But looking at sixty years of numbers told a clear story.

To help understand why water levels had dropped so much and so quickly, Clem turned to the county government. Clem remembered saying to the county officials, "We're losing water. We are a lot dryer in the dry season, we're seeing the vegetation change, we've lost our wood storks in most years. We need to figure out what's happening. We need to narrow it down and figure it out."

It turned out that Collier County already had the answer. "We start putting the puzzle pieces together," Clem recalled. "And they're like, 'Oh, yeah, well, there was this canal improvement project that happened. And you know, people just south of you, their road was flooding in the rainy season. So we made those canals work better.'" The county had changed the shape of the canals to hold more water, and they had also reconfigured bridges and culverts so that more water could flow underneath roads—preventing the roads from being barriers to water movement. "They improved flow control for the residents down there, which sucked water out."

— – – —

The area south of Corkscrew that benefited from the deeper canals lies in a development called Golden Gate Estates, with a long and sordid history of crooked land dealings, like the many that have shaped south Florida. In the 1960s, two brothers from Baltimore, Leonard and Julius Rosen, formed a company called Gulf American, which planned the biggest subdivision in the world—about 114,000 acres—in the middle of a swamp. The Rosens had gotten rich hawking a bogus anti-baldness tonic, and as journalist Michael Grunwald wrote in his book *The Swamp*, they "could see that shivering northerners yearned for a piece of Florida the way bald men yearned for hair."

Suburban housing near Naples blocks wildlife corridors and alters water levels in surrounding wetlands. *Carlton Ward Jr.*

The Rosens produced fancy brochures touting elegant living amid tennis courts and fine dining, even though not even a sewer line had been built. As Grunwald described it, no tactic was too unethical in selling the swamp. "Salesmen told veterans the firm was affiliated with the military. They relocated lots without informing buyers, sold lots fifteen miles from the Gulf as 'waterfront properties,' and preyed on the senile and feeble. They drove prospects who insisted on checking out their land deep into the swamp, then threatened to make them walk home if they didn't sign a contract." The company made close to $115 million flipping lots in the wetlands, and after eventually pleading guilty to "deceptive sales practices," one Rosen brother ended up appointed to the state's land sales board (for which he later awarded the then-governor with a job as a consultant).

The southern section of the Estates ultimately got infrastructure—canals, roads, drainage. But it didn't get houses. That's the portion that is now part of Picayune Strand, the state restoration project, the first of sixty-seven Comprehensive Everglades Restoration Plan components to break ground, back in 2007. The north Estates is about half-developed, with houses and some amenities but notoriously poor community services (bridges, sewers, parks). Parts of it look like any other Florida community, ranch homes neatly laid out along grids of roads and canals. But the farther east you go, the less developed it is. And today, the undeveloped lots are mainly in the lowest-lying areas. Local conservationists are targeting these undeveloped acres as a way to help restore habitat for wildlife like Florida panthers and black bears, which need space to roam and corridors to move between protected natural areas, and also to help hold water for the birds and other critters that thrive in the wetlands. Florida sees a lot of overlap between its remaining wetland ecosystems and wildlife corridors, in part because the higher ground was developed first. The deepest wetlands are relatively

less covered in roads and human communities, so those are, now, both the most important places for wildlife movement and the last available sites for building a home or a Home Depot.

One steamy July morning, I met up with Bradley Cornell at a Panera in Naples to learn more about the complicated relationship between Corkscrew's delicate hydrological balance and the subdivisions that abut it. Cornell has lived in the Naples area for thirty years—he moved in the early 1990s for a job playing trombone in the Naples orchestra. In his free time, he volunteered on various conservation projects, and eventually he realized he had found his calling. Cornell is the southwest Florida policy associate for both the state and local Audubon organizations, Audubon Florida and Audubon of the Western Everglades, a job that requires a seriously in-the-weeds knowledge of the ecology and the politics of the area.

The Corkscrew Swamp Sanctuary sits in Collier County, a fast-growing, highly conservative area in the southwestern part of the state, where Donald Trump won three times. Despite its anti-regulatory politics, though, Collier County passed, in 2002, a referendum that increased local property taxes for ten years to enable the county to purchase and preserve land. From 2004 to 2013, the county bought more than 4,300 acres of land for conservation. In 2020, at the same time the county voted by more than 60 percent to return Donald Trump to the presidency, it voted overwhelmingly—76 percent—to reauthorize the land-preservation tax for another ten years.

I asked Cornell why, in this conservative and tax-averse corner of the state, he thought residents had voted to tax themselves to buy land for conservation. "People want to see this stuff protected," Cornell said. "They don't want to lose all this stuff. They also want to live here. So that's a little bit of a paradox. But I think the public support for conservation, even on a national scale, outstrips the hyper-partisan politics of blue

versus red. It's something for everybody. And everybody has an interest in seeing this succeed."

After we finished our coffee and pastries, Cornell drove me to see one of the projects that the voters of Collier County were supporting. Since 2021, Audubon has partnered with the county to purchase lots in a band of green space called Horsepen Strand through the taxpayer-funded program. It's a key link that can provide safe passage for panthers and other wildlife, and also help retain water, and it's a conservation priority for both Audubon and the water management district, which needs a place to put floodwater so the roads don't become submerged. "Let's put it in a cypress slough, what a novel idea," Cornell said. It's the human condition: Destroy natural systems, suffer the consequences, engineer a solution that approximates what nature did or attempts to recreate what we unwisely demolished. The county program had so far purchased around 30 lots, out of 300 total needed to assemble the wetland corridor.

Horsepen Strand sits in the part of Golden Gate Estates that's closest to Corkscrew Swamp Sanctuary, where many lots remain undeveloped. They're just too wet. On a map, the neighborhood appears as a peninsula of development jutting into a horseshoe of protected areas—the Audubon sanctuary and a string of other preserves run by a local land trust. Sadly, between the subdivision and the protected areas is a two-lane road, where at least eight panthers and six bears were killed by vehicles on a thirteen-mile stretch, from the early 1990s through 2019.

We drove through the streets around Horsepen Strand, in the center of the subdivision, where many driveways, and

A panther makes its way down a flooded logging trail in the aptly named Florida Panther National Wildlife Refuge, which protects important habitat to the east of Golden Gate Estates. *Carlton Ward Jr.*

sometimes entire front yards, were underwater. Some streets looked more like rivers, with egrets and herons standing in the water. This was not exceptional flooding; it was just rainy season in the swamp, and it was hard to tell what was a year-round pond and what was a flooded lawn. Either way, it seemed like an ill-advised place to build a house. The homes that had been built were up on berms known as fill mounds.

Cornell said every new concrete or asphalt surface brought the system further out of balance. "It's like putting bricks in a bathtub," he said. "Eventually, it overflows, right? You lose your storage capacity." The green space would help hold water that would both prevent flooding and percolate through the system instead of dumping into canals that flushed it out to sea.

We drove past an elementary school and turned down another street, past a few more flooded driveways and a good number of For Sale signs. Finally, the road dead-ended at a canal. Just below the surface, water was flowing into the canal from a large black corrugated plastic drainpipe. Cornell said it was water leaking out of the sanctuary.

"You know, for forty years, our swamps behaved a certain way," he recalled, as we stood at the edge of the canal. The water filled and drained from the different elevations and pools in relatively predictable cycles, save for that little bit of year-to-year variation. "And all of a sudden, in 2000, they didn't anymore."

Clem's research, Cornell said, "told us what we should have all realized—that canals were really effective at pulling the water out of Corkscrew Swamp Sanctuary." The canals were built to keep the residents of Golden Gate Estates from having rivers where their roads should be. But the fact was, their roads shouldn't have been there at all. None of it should have. And now, the canals were draining water from the sanctuary as well. "The sanctuary has so much development on its heels, and downstream"—where we were standing—"everybody's

screaming, 'Get this water out of here!' So they're sucking it out." The canal enhancements in 2000, Cornell said, "pulled out the last plug in the bathtub, if you will, and all the water now, four out of five years, drops below ground." Standing there watching the water flow into the canal, it seemed particularly sad that the wood storks were suffering at least partly due to this cheap plastic hose. But the problem was clearly larger than just one pipe.

Clem had initially been worried about telling officials from the water management district—the very people who'd built the canals—that their canals were the problem. But they were "really concerned, they said, 'We need to get together to fix this.'" One promising solution seemed counterintuitive: build a wall. A clay wall, specifically, buried several feet underground, perhaps three miles long and a few feet high, on the south side of the swamp. During the wet season, the water would simply spill over the top, as though the buried wall wasn't even there. But in the dry season, it would trap water in the system, keeping it in place when the birds need it most.

— – – —

On the day when Clem and I found ourselves briefly stuck in the rut near the alligators, she had generously agreed to show me around a part of the sanctuary not open to the public. As we drove through the more remote parts of the preserve, she pointed out some of the ecological interactions that were taking place all around us. Her background is in community ecology—a subfield of ecology concerned with the interactions between species living in the same geographical area. Because of this, she views Corkscrew through the lens of food webs.

In an area of flooded prairie, we watched an anhinga standing in the water with a catfish in its beak. The elegant bird seemed to be trying to figure out how to consume the fish. To Clem, this indicated that the catfish was a non-native species. "There's

a couple species of armored catfish from South America, and they're really tough for these guys to handle," she said.

We stopped to observe a green heron on a tree branch and caught a glimpse of a limpkin flying past. "They're neat birds," Clem said of the tropical wetland species, which is related to cranes. Limpkins, so named because their gait apparently looked like a limp to European settlers, have beaks that are bent at the end in just the right configuration for eating apple snails. Florida is currently home to five species of apple snail, but only one is native. Limpkin numbers were "crazy low," Clem said, "until this non-native apple snail came in. They love them." It was nice to hear a hopeful wildlife tale, even if it involved an invasive species. Along the edge of the road, I could see large white snail shells. The Everglade snail kite, an endangered bird of prey with a similarly curved beak that lost most of its habitat as humans drained and paved the Everglades, had also rebounded after another non-native snail arrived, Clem said.

"It's a supercool example of rapid evolution," she said. The snails "were too big for the snail kites to eat, especially the juveniles. Over the course of ten or fifteen years, they were able to measure a rapid evolution of snail kite beaks to be able to eat these snails. Just over a couple of generations' time. And now they're doing really well."

Elsewhere along the road, we paused to watch a raccoon and a softshell turtle having a strange moment together at the base of a tree. The raccoon was sniffing near the turtle's back end, hopping around eagerly, almost as though it was trying to figure out

An Everglade snail kite comes in for a landing with a tasty meal at Florida's Lake Okeechobee. This endangered raptor, which dines on apple snails, lost much of its habitat when the Everglades were drained. Restoration efforts have helped the kite rebound and so has the presence of a non-native apple snail that has colonized the ecosystem. *Sydney Walsh/ National Audubon Society*

what the turtle was. "He's like, 'You laying eggs or what? I can smell them. But what are you?'" The turtle eventually waddled off in one direction and the raccoon trotted away in the other.

We passed an area clogged with Carolina willow, a native species that has grown out of control, capitalizing on changes in hydrology to expand its range. It's an important plant for birds, hosting caterpillars and other tasty insects. But across Florida, it has recently taken root in thousands and thousands of additional acres of marshland, crowding out native grasses, because it has deep roots that can tap water further into the dry season. Because it has larger leaves, Carolina willow loses more water through transpiration than the grasses it is replacing, so as the climate warms, it is using up even more of the available water. It also is shrinking the habitat available to birds, which can't penetrate the dense vegetation to feed, and other animals, like deer and bears, which can't pass through. The thickets of willows, despite being native to the ecosystem, become barriers to connectivity when the ecosystem is out of balance. And in Corkscrew, they are taking over critical wading-bird foraging habitat. Disconnecting the hydrological cycle was transforming the whole ecosystem.

The willows transpire so much water that they create a humid microclimate that is fire-resistant, which means burning them isn't an option for keeping them in check. (This fire resistance is also an issue for prescribed burns, which are necessary both to reduce the risk of catastrophic wildfire and to keep other vegetation in balance.) Corkscrew staff tried aerial spraying with an herbicide. This killed the willow, but then nonnative plants sprouted up in its place. And with the willows dead but still standing, it was impossible for anyone to get in to deal with those plants, which further upset the balance of the system. Then the staff tried a new approach: mechanically shredding the willow. In the dry season, they bring in machines on treads with big mulching jaws. The machines grind the willow

and leave the woody material in the marsh, to break down when the water levels rise. The tactic has worked. But it's a slow and expensive process, and they can only manage about 250 acres a year. So far the sanctuary had cleared about 1,000 acres, which has had a "tremendous local impact," Clem said. But there were another 1,000 acres to do, and the willow keeps expanding. The aim is to reduce it enough to get fire back into the system and control its further expansion that way.

We drove through an oak hammock on higher ground, forested with hardwoods and palms. We passed another, more open area, where in the 1970s the sanctuary had experimented with building fish ponds to feed the wood storks. The plan never worked—the ponds wouldn't hold water at the right times, the storks quickly cleared out all the fish—and the area is now a favorite hunting ground for panthers, which prey on raccoons, opossums, armadillos, and feral hogs that roam the area. It's also the place where Clem feared they would one day find a Burmese python, the invasive snakes that are deadly to native birds and lots of other species. "The python hunters have told us pythons will love this." If it wasn't one thing, it was another.

Burmese pythons wreak havoc on pretty much any ecosystem they colonize. They have moved into and through the Everglades in part using the same human-created corridors that disrupt so much other animal migration. "Pythons are making their way across the state because of this matrix of corridors that we've built, in canals and levees—we've built all these animal superhighways for pythons." They hadn't yet found pythons inside the sanctuary, though a few had been killed on the road that leads in. It was only a matter of time.

"Seeing raccoons out like that during the day is a great sign," Clem said. "Because if you go down to Everglades National Park, you won't see raccoons anymore. Like, they're just gone. No raccoons, even in the campgrounds. There's hardly any white-tailed

deer, there's no possums. There's no armadillos. All because of the python." The sanctuary's biologists were working with University of Florida researchers on a "python response plan," for when the inevitable happened, and also learning to try to spot the snakes as quickly as possible. "They have a really low detectability rate. So we probably will have them long before we ever know we have them. We're trying to figure out, how do we overcome that? How do we make sure we know as soon as possible so we can start trying to remove them or manage them?" The decommissioned fish farm was at the top of the monitoring list.

We drove out into open marshland and stopped to watch an American coot in the water. This was the habitat that wading birds preferred, and it had recently been choked by willow that grew twenty feet tall. Now it was a sparkling wet meadow again. A great blue heron flew past us. Clem said this was a hotspot of herons, egrets, and spoonbills in the dry season, and that there had been a flock of a hundred white pelicans here for a while earlier in the year. I asked if wood storks liked this area, too, and Clem said the storks liked to nest in cypress trees with standing water beneath them—particularly water with alligators in it. Alligators and wood storks, it turns out, need each other. The alligators hang out underneath the trees, waiting for any regurgitated fish—or even a chick—that might fall out of the nest. But the alligators also act as security guards, protecting the storks from predators like raccoons, which prey on both chicks and eggs. "When we over-dry the swamp and we lose the standing water underneath the cypress trees in any part of the dry season, then raccoons can get up there," Clem said. "We've had raccoons

American alligators are an important part of the ecosystem in Big Cypress National Preserve, near Jerome, Florida. In exchange for protecting wood storks from raccoons, they reserve the right to eat anything that might fall out of a nest. *Sydney Walsh*

clear out part of a wood stork rookery, which really shouldn't happen if our hydrology was right." No water means no alligators, too many raccoons, and ultimately no wood storks.

Water was the connective substance holding the whole ecosystem together. The way it flowed and percolated through the landscape governed, directly or indirectly, every interaction—between predators and prey, between plants competing for space, between the species that had existed there for millennia and those arriving because of an altered world. Like so many other creatures, the wood storks that once thrived at Corkscrew were fundamentally linked to the water cycle here—and fundamentally threatened by how it had changed.

The day after Clem showed me around, I took a slow solo stroll on the wooden boardwalk that runs through the public part of the sanctuary. I arrived at 8:30 a.m. and had the whole place to myself for a good hour, during which the only other creature I encountered was a raccoon. It was walking toward me on the elevated trail, wary but seemingly unafraid. I stopped, curious how this interaction was going to go down. The water beneath us was several feet deep, which was why the raccoon was using the walkway. Again, human infrastructure was functioning as a wildlife corridor. I wondered how the boardwalk itself had changed the system in some tiny way. It was only a few feet wide, and neither the raccoon nor I had anywhere else to go. I stood still as the animal waddled nonchalantly toward and then past me, and then each of us continued our morning walks. It turns out running into wildlife on the boardwalk is not an uncommon occurrence—visitors have even encountered panthers. One viral video from 2016 shows a panther trotting along the walkway toward the camera, then—clearly freaked out—skidding around a bend in the path as it races off.

The Everglades was full of examples of connectivity, both visible and invisible, linear and non. You needed corridors of

green space like Horsepen Strand, and natural areas that could retain water like bowls, and not so much asphalt and concrete that there was nowhere to store the water and no way for it to slowly seep through the landscape. But there was good news: Clem's hydrology work had gotten traction, and the South Florida Water Management District—the state entity responsible for flood control, water supply and quality, and much of the Everglades restoration—was taking up the cause. They were interested in hydrological restoration strategies that could help with wildlife conservation, flood control, drought resilience, and wildfire prevention, and had invested hundreds of thousands of dollars in modeling efforts to design solutions. "The big positive thing is, they've made the commitment to fix it," Clem said. "The ball is rolling forward, they've budgeted some funding. But it's a big project. It's a mini-Everglades restoration, and there will be lots of stakeholders."

One seemingly straightforward idea was to link some of the homes downstream of Corkscrew Swamp Sanctuary—in the area I drove through with Cornell—to the city of Naples's sewer system. These houses were still on wells and septic tanks, which meant that raising groundwater levels could end up flooding the septic drain fields and polluting the water. Converting to a sewer system would enable groundwater levels to rise with no issue.

Clem and Cornell seemed cautiously optimistic that the hydrology could be repaired and the wetlands could hold water again. In the meantime, they hoped the wood storks could hold on. I did too. There were so many ways to break the hidden but crucial links holding ecosystems together, so many examples where seemingly minor actions in one place reverberated far away. But those seemingly minor actions worked both ways: They could also help fix things. I'd seen it in Mexico, in Italy, in Costa Rica, and I was about to see it in Kenya. The most crucial ingredient, always, is people who care.

CHAPTER 10   *Kenya*

The Time Is Now: Elephants and the Gullies of Kenya

The regularly used airstrip outside the town of Archers Post, in northern Kenya, was closed after being damaged in heavy rains, so my flight landed at an even tinier dirt airstrip on the other side of Samburu National Reserve. From there, the road headed south across community-owned grazing lands to the western gate of the reserve, along the Ewaso Ng'iro River. It was just a week into 2024, and January is normally the start of the dry season, but the rains still hadn't fully subsided. The reserve, a protected area about the size of Washington, DC—part of a much larger swath of adjacent conservation lands—was a study in green, its plains and hillsides covered in a tangled tumult of vegetation.

Previous Spread: Young elephants frolic in Samburu National Reserve, in central Kenya. These highly intelligent mammals migrate seasonally across a broad ecosystem here, between Samburu and the plains of Laikipia, following longstanding, culturally shared routes. But more and more, they are bumping up against new infrastructure and development.
Michael Nichols

Almost as soon as we crossed the reserve boundary, we were among elephants, ranging in size from months-old babies to small and spunky adolescents to towering matriarchs and giant bulls. They lumbered about on their baggy-skinned legs and round feet, they snoozed, they roughhoused, they nursed, they waded in the river, they doused themselves with dirt. In the midday heat, they stood in the shade of acacia trees and lazily plucked up wads of greenery with their trunks, then swung it up into their mouths. They were unbothered by the presence of the Land Cruiser I rode in; they wandered right up to the vehicle, some barely more than a trunk's length away.

These were salad days for the Samburu elephants. It had been a long, hard, deadly two years of extreme drought, when starving animals succumbed in the desiccated landscape. Between 2021 and 2023, 1,260 elephants died in the broader Samburu-Laikipia region, at least 563 from "natural causes"—most likely hunger and thirst from the drought—and a good many of the rest at the hands of humans fed up with desperate animals crashing through fences and stealing the meager grasses that herders needed for their also-starving livestock. But the rains had finally come in November, and kept coming, and the region now was as verdant as anyone could remember seeing it for years. The elephants were fat and healthy, and it was magical to watch them enjoying a rare period of relatively easy living.

At a bend in the Ewaso, across from an old safari lodge destroyed in another flood and now given over to baboons, sits a small field camp where conservationists have been studying and helping protect elephants since 1997. Researchers there can identify more than 600 living elephants by sight, based in part on detailed notebooks containing sketches of the animals' ears. They've given family groups thematic names—Flowers, Royals, Hawaiian Islands—to help keep track of the individuals within each group, recording each birth and death. And they've

also been fitting elephants with GPS tracking collars, amassing three decades of movement data that helps inform conservation planning. At the field camp, wooden racks set between the river and the open-air dining area hold scores of used elephant collars, giant leather straps outfitted with rectangular boxes—a museum of GPS-collar design throughout the ages.

Iain Douglas-Hamilton, a Scottish zoologist, started Save the Elephants (STE) in 1993, after studying elephants in East Africa beginning in the 1960s. Much like Jane Goodall gave humans intimate access to the world of chimpanzee behavior and culture, Douglas-Hamilton did the same for elephants. He also brought the poaching crisis to the world's attention and helped spur global ivory bans and conservation initiatives. In the 1970s, Douglas-Hamilton kept tabs on elephants from the air, piloting a tiny plane and, as several people who flew frequently with him recounted to me, often swooping down sharply and nauseatingly to follow a group of animals on the ground. These surveys laid the foundation for elephant movement data that showed just how crucial it was to protect the corridors. Later, Douglas-Hamilton pioneered the use of GPS tracking collars on elephants.

Years of elephant movement data shows distinct elephant highways, routes the animals have used throughout history to access food and water. Today, though, those routes are disappearing. By building roads and rail lines, putting towns and cities in the midst of elephant roads, and erecting fences that make it impossible for the animals to cross, humans are cutting off the elephants' ability to move across the landscape. As these highly intelligent mammals follow their longstanding,

A radio-collared elephant in Samburu, Kenya. Save the Elephants has amassed three decades of movement data on the elephants that live in this ecosystem. This information helps inform conservation planning and is particularly important as Kenya experiences a period of rapid urbanization. *Lisa Hoffner /NPL/Minden Pictures*

culturally shared routes to reach the resources they need seasonally, they bump up against new infrastructure and, often, few desirable alternate pathways. This raises the likelihood of human-elephant conflict, which in Kenya has now become the largest cause of elephant deaths, just as the herds were finally recovering from years of a devastating poaching crisis—a "holocaust," as Douglas-Hamilton has referred to it.

In northern Kenya, many of the elephants that graze in Samburu National Reserve are migrants. They follow the available food, moving between vast tracts of protected areas, often traveling hundreds of miles each way across community-owned lands—the matrix—with some places intentionally set aside for livestock grazing and wildlife movement and others not. Samburu was lush when I turned up in January, but within weeks it would begin to dry out. Water levels would drop, nourishing food would be harder to find, and many elephants would begin to venture elsewhere. Hundreds of them would migrate west, making their way along various well-worn routes and then funneling through a narrow channel between steep hills, near the settlement of Oldonyiro, on their way to the plains of Laikipia. There, acacia trees would be dripping with the tasty, nutritious seed pods the elephants crave.

That area around Oldonyiro is a crucial link on the elephants' journey. It's also a study in just how quickly things in Kenya are changing, and what's at stake.

— – – —

Kenya is a country in rapid transformation. Under a program launched by then-president Mwai Kibaki in 2006, the country has been striving to become a middle-income economy by 2030. A central component of this plan, known as Kenya Vision 2030, has been infrastructure improvement—including a goal of paving 10,000 kilometers (more than 6,000 miles) of roads and modernizing the existing railway system, built by the colonizing

British a century ago. Beginning in 2014, as an early initiative of China's Belt and Road Initiative, the Chinese government funded 472 kilometers (293 miles) of new train tracks, called the Standard Gauge Railway (SGR), stretching from Nairobi southeast to Mombasa, on the coast. That train line cut the journey by rail from about twelve hours to four, but it also bisected two national parks with tracks. Next up is the expansion of the railway in the other direction, and the construction of a new six-lane expressway, parallel to the SGR, where a two-lane paved road currently moves roughly three-quarters of all goods imported to East Africa.

Where roads go, so do development and sprawl. Kenya sits at a crossroads, with one path leading to sustainable growth and the other to massive collapse of the country's iconic wildlife populations. Ensuring that those animals can safely cross the matrix is a daunting, essential conservation challenge, with implications far beyond the country's borders. As one of the leading economies in Africa, Kenya is under scrutiny as a model for how to develop and raise living standards without sacrificing biodiversity. Getting it wrong could have wider, catastrophic, implications.

— – – —

On a sunny afternoon, I set out for Oldonyiro—about four hours' drive from Archers Post along narrow, rutted dirt paths—with Benjamin Loloju, who is Samburu and grew up herding goats in the surrounding lands. He told me he thought he was around thirty-two or thirty-three years old, but he didn't know for sure since no one in his community recorded births. One of eleven children, he was the only member of his family to attend school; the literacy rate in the area still hovers around 30 percent. He excelled at school and ultimately earned an "elephant scholarship" from Save the Elephants, part of an initiative that helps fund schooling for children in pastoral communities coexisting with elephants. A new dirt road was under construction at the

time, a big step forward for development. Roads meant more settlements, more permanent structures, and more potential for escalating human-wildlife conflict. Douglas-Hamilton wanted to work with the community to promote coexistence.

With funding, Loloju attended boarding school in Nairobi. "I could not believe it the first time I arrived in that city," he recalled, laughing at the memory. He earned straight As and made the list of the top one hundred students in all of Kenya. He went on to study geospatial engineering at the University of Nairobi, and then earned a master's degree in the UK. When I met him, he was beginning to think about PhD programs to study soil erosion, a particular problem in this region. Loloju built a home on a hill outside Oldonyiro, returning to his homeland to help create safe and sustainable pathways for elephants.

We bumped east across vast plains of communal grasslands, occasionally getting stuck in the sand crossing ravines or dry riverbeds. We skirted dramatic hills and rock formations where baboons scampered. We passed zebra and giraffes, tiny dik-diks skipping through the bush, grazing impalas, and the occasional temporary human settlement, or boma, with huts constructed of acacia wood. Every so often we passed a lone herder walking with his goats, or a child carrying containers to fetch water. But mostly we drove for hours across expansive landscapes with no permanent human structures. A few hours into our journey, we stopped where a trail crossed the road. Loloju hopped out to show us a square concrete marker placed in the dirt. "Livestock + Wildlife Corridor," read a metal sign embedded in the concrete.

Benjamin Loloju, corridors manager for Save the Elephants, stands in an erosion gully in Oldonyiro, Kenya, near where he grew up. Elephants making their seasonal migration between Samburu and Laikipia must cross through this area, navigating the gullies—caused by a mix of grazing patterns, climate change, and physics—as well as new roads, buildings, and lots and lots of fences. *Jane Wynyard/Save the Elephants*

Based on the data from the GPS-collared elephants, Save the Elephants and Wyss Academy for Nature, a Swiss organization, have worked with communities across hundreds of miles of northern Kenya to delineate these corridors, so that any future development, should it occur, will protect at least this pathway for the animals. They have installed more than 200 markers in the ground, with community support. The corridor demarcation helps ensure a route for herders to move their livestock, as well as a route for wildlife. "This has been good," Loloju said of the effort. "People, communities, have agreed to say, 'Good, demarcate it.'" The group is now working to get these routes enshrined in national law, and also to extend them farther east and south, in Laikipia. But Oldonyiro represents the thorniest situation. "We already think maybe we have been late for Oldonyiro. We should have come ten years earlier," Loloju lamented. "But now, this is the time."

In the late afternoon, we came to an area of more concentrated settlement than we'd seen all day, with permanent structures—buildings made of cinderblocks and concrete—beginning to appear along the roadside. This was the edge of Oldonyiro. "It is such a tricky, tricky situation here," Loloju said. "This is a really, really big lifeline corridor." From Samburu to Laikipia, it's essentially one big migration area for elephants, and so many other animals, that need to move with the seasons. "But here"—in Oldonyiro—"is the thread; here is the heart of it. If we close off here, then we will cut off those elephants moving, and other wildlife. We will cut off those movements. And for their survival—for their *survival*—it is so, so important to have this connection." Without it, many more elephants could die, from lack of food or at the hands of angry humans.

On a map of the elephants' movement data, big spaghetti-like blobs of red lines, showing the paths of individual GPS-collared animals, indicate "relaxed" movements—elephants eating,

drinking, frolicking, resting, maybe mating. They're on their way someplace, but they aren't in a crazy rush to get there. Imagine you're on a long-distance drive, and this is where you stop for a leisurely lunch, spend the night, maybe visit a local attraction.

Right near the town of Oldonyiro, though, that spaghetti straightens into a single noodle. Here, the data makes clear, the elephants are not hanging out; they aren't dawdling. They're just trying to cross to the next safe space. This is where, on your road trip, you've reached an unpleasant stretch of highway, or maybe you're just trying to power through and put the miles behind you.

On the same map, yellow circles indicate human settlements. The biggest cluster of these intersects with the elephants' no-dawdling spaghetti strand exactly in the town of Oldonyiro. For centuries, perhaps millennia, elephants crossed the landscape here. But now it's a newly bustling town center, with building after building rising up along the dirt road. An area composed previously of pastoralists' bomas is morphing into a town. "All these are new," Loloju said as we drove through the center of the development, where one-story concrete buildings—butcher shops, basic grocery stores, mobile phone kiosks—were under construction on both sides of the street.

He pointed out a handful of wooden structures that dated back to his childhood. Other than those few buildings, the rest were recent additions. "This expansion is heading that way," he said, pointing south. "And it will go all the way to where I live, in the next very few years. Everything is new and they're still building—and so the line is approaching the corridor that the elephants are using at the moment."

Next Spread: Elephants crossing the Ewaso Nyiro River, Samburu National Reserve, Kenya, near a field camp where researchers have studied and worked to conserve elephants since 1997. Many of these elephants make the seasonal journey from here to Laikipia, passing through Oldonyiro on their way. *Courtesy of Frank af Petersens / Save the Elephants*

Barred from their historic route because the new buildings now blocked their path, the elephants had forged a new route south of town. The latest map clearly showed their shifted movement. You could even see from the GPS tracks that elephants had tried to use their traditional route but hit fences or other development, made U-turns, searched for a way around. A few family groups had found this new southern route, and "everyone else I think will smell, and they know, this is now the route," Loloju said.

"Elephants are very adaptive," elephant biologist Lucy King, who grew up between Kenya, Somalia, and Lesotho, had told me before my trip. King works as director of STE's Human-Elephant Coexistence Program and pioneered the use of "beehive fences" as a way to prevent elephants from raiding croplands in southern Kenya and elsewhere. Elephants want nothing to do with the bees, so putting hives on fences can help keep them away from crops. "They are quite capable of moving and changing where they walk, if they can find another route. They don't mind being a bit flexible." But there have to be available options.

As it turns out, the elephants' new pathway south of town is also critically endangered. Two years earlier, the county completed a land conversion here, transitioning from community-owned land to private title deeds. People now owned their individual lots—and many were building permanent structures and enclosing their property with wildlife-proof fencing, not unlike in the American West. "These elephants will come and say, my god, this fence was not here a month or two ago," Loloju said.

We pulled over at a point where the elephant corridor crossed the road, and we strolled a hundred meters or so down the trail.

Save the Elephants team members check on a beehive fence in Tsavo. These fences are just one tool the organization has developed to help protect both elephants and nearby communities. *Jasper Scofield/Save the Elephants*

Dried mounds of dung dotted the rust-colored dirt. "This is someone's plot," Loloju said of the ground we were standing on. "It's just not built yet." He pointed around us in various directions. "Here is someone's plot, it's a title deed for someone. On this other side, it's a title deed. On this other side, it's a title deed. So in a few years on, all this could be built up." Once that happened, the route would be completely cut off to elephants— unless Loloju and some local helpers could convince the landowners around the corridor to set aside narrow sections of their properties, just twenty or thirty yards, even, for wildlife—like a conservation easement. It's unclear how far the animals might walk in a fenced-off corridor that narrow, but it represented the only option.

Across the road, we could see the future. The newly titled landowner there had already built a large home, surrounded by a six-foot-high wire fence. One section of fence was missing; elephants had knocked it down recently as they tried to follow the route they had taken just months earlier, and the owner had piled acacia branches to temporarily fill the gap. Loloju had been speaking to individual property owners—a meeting with several dozen of them was scheduled for the following month—and this owner had verbally committed to build his fence leaving thirty meters for elephants. But then he built the fence far closer to his property line. And because the fence now sat exactly in the path that the elephants had last walked, they had returned, come too close, and damaged it.

This wasn't the only problem. Part of the reason the elephants had stayed so close to the fence line was that it represented the least dangerous avenue for travel. All across the area, giant erosion gullies, some large enough to drive a car through, are opening, spreading, and multiplying. It's a wicked problem in which grazing patterns, climate change, past land-use mistakes, and sheer physics are combining into a quagmire—both

literal and metaphorical. Livestock grazing removes vegetation, which loosens the soil—and then when the rains come, the pounding water erodes the bare dirt. "It starts like a little path," Loloju said. "And over the years, it just expands." With nothing to hold the soil in place, the gullies grow and grow. Many are far too wide now for a human to jump across, and more than six feet deep—large enough for baby elephants to fall into.

We walked for an hour across the plain, peering over the edge of gullies, trying to follow the elephants' trail. Some gullies we could maneuver around, some we could hop over, some were so wide we had to scurry down one side and back up the other. It was hard to imagine a family of elephants making this journey; even harder to envision how they would make the trip if fences left only a narrow corridor and then that was swallowed by a sinkhole.

This wasn't an abstract exercise of thinking about some hypothetical elephants or picturing a theoretical route. I had seen the exact individual elephants that would soon be heading this way—seen the adults close ranks to keep babies safe in their midst, watched one frisky young sibling try to wake his sleepy brother from a nap, seen their wrinkles and eyelashes and how the dirt clung to their tusks. These were *the* elephants that, in the coming weeks or months, would arrive in this very spot, needing only to safely pass through on their way to the next place where they could have plentiful food.

— – – —

Loloju had turned to the Kenyan Army for help, hoping they might have some ideas for how to fill in the gullies, or change the grading to make it easier to cross some of the worst ones. He was awaiting their reply. But it was a vexing engineering challenge. Each time the rains came, new sections of ground would collapse. In some gullies you could see wire-mesh cages filled with rocks—valiant attempts at slowing down the

rushing water—wedged at awkward angles where instead the water had dislodged them. Planting cover crops might help hold the soil in place and prevent further erosion, but that would require setting the area off-limits to grazing—not an option in a place where livestock are the community's lifeblood.

Near Oldonyiro's small school, Loloju took me to a spot where, as a child, he would step across a small crack in the ground. Now it was a gaping gash in the earth, maybe five feet deep and several feet across. "When I was a young boy, when I first came to this school, oh my God, it was a little tiny thing that I'll just go over with a step," he remembered. "Now, there is no way I can cross that." In addition to trying to find an engineering solution, he was also working to convince landowners to help preserve the corridor—and both those things in themselves were full-time jobs. But he was also trying to secure a backup plan: yet another, alternative, route for the elephants should all else fail. The regional terrain and the patterns of development had left open just one other potential corridor, this one to the north of town. There, the land was not yet privatized, though that was coming. Before that happened, while the land was still community-owned, STE was trying to delineate a wildlife corridor.

Towns, fences, giant gaping sinkholes in the ground—these were the things the elephants were up against. But there was also the issue of anthropogenic resistance. I asked Loloju why people in the community might want to work with a conservation organization to protect elephant corridors. He had a list of reasons at the ready. Some people take to the idea of delineating corridors away from schools or houses, to keep people safe—children in

Mpayon Loboitong'o is a leader of the Mama Tembos, a group of Samburu and Turkana women who help monitor the elephants who move through their communities. She proudly displays an issue of *National Geographic* titled "Women: A Century of Change." *Jane Wynyard/Save the Elephants*

SPEAKING
OUT
TAKING
CONTROL
CHANGING
DESTINIES
SHAPING
THE
FUTURE

Women around the world are making their
voices heard in government and
their communities, moving many closer
to gender equality.

BY RANIA ABOUZEID

PHOTOGRAPHS BY LYNN JOHNSON

particular. Others recognize that money from wildlife tourism can help bolster community conservancies, funding schools and clinics and other facilities. Some are persuaded merely by empathy for the animals. And still others like the idea that there will be people monitoring the corridors, keeping an eye out for trouble. STE runs a program called Mama Tembos, meaning Elephant Mothers (modeled on a lion guardian program, Mama Simbas, run by another conservation group), that employs women from the Samburu and Turkana communities alongside the corridor to monitor it.

I met with four of the most experienced Mama Tembos from communities near Samburu National Reserve one morning near Archers Post, where the A2 highway, a paved thoroughfare that runs from Nairobi to the Ethiopian border, about 300 miles north of Archers, intersects the elephant corridor. There, community lands known as group ranches connect with several national reserves. Each morning, the Mama Tembos walk a kilometer or so down their assigned corridor, checking for animal footprints. They also get reports from the community—about a sick elephant or an animal that is stuck somewhere and needs help, or if a large group of elephants passed through in the night. They report to their local conservancy warden if they find that people are settling within or close to the corridor. And they also function as community educators, passing along news of approaching wildlife so that people can stay out of their way.

Wearing the heavy beaded collars that symbolize their tribal and marital status, the women spoke about their work, through a translator. I asked what challenges they faced, and Jerenica Loole, who had been doing the work for about five years, said that occasionally when elephants cause trouble—as they did during the drought—the Mamas get blamed, with angry community members referring to *"your* elephants." But that was the only problem she mentioned. She said that understanding

what an elephant is had been an important element of the work. Initially an elephant was just seen as a big, gray creature that threatens you, she said. But now they saw the animals differently—as intelligent beings that had families and cared for one another. She also said that providing employment opportunities, especially for women, had helped change attitudes toward wildlife more broadly.

— – – —

While northern Kenya's dry climate makes it virtually impossible to do any sort of farming, southern Kenya is strikingly different. There, conservation has butted up against an influx of land sales for large-scale agriculture, which threaten to obliterate wildlife habitat and corridors in the matrix, as well as further erode Indigenous lifestyles and livelihood. Land privatization in a place where agriculture is possible and profitable has led to rising land values, subdivision, fencing, and land-grabbing— in which local landowners sell to outside investors who swoop in for a bargain. A recent and highly publicized controversy shows the scope of the threats: After a vast area of formerly communal grazing lands near tiny Amboseli National Park was subdivided and privatized, one newly titled owner sold his 180 acres to a Nairobi-based avocado farming company. That company, KiliAvo, was able to buy the land for agriculture despite it being in a designated wildlife corridor in the heart of an elephant migration route.

The local Maasai community and some conservation organizations managed to successfully fight the farm—thanks in part to GPS-collar data showing that it was square in the middle of an elephant corridor—but not before the company had cleared and fenced the land. Samuel Ole Kaanki, head of a group representing Maasai landowners in the area, called the avocado farm "a disaster" in an interview with the UK newspaper the *Independent.* "It is located in one of the critical migratory

corridors that are key to wildlife," he told the paper, adding, "This will result in massive human-wildlife conflicts."

It's easy to understand why someone would be tempted to sell their land. It brings in "a huge amount of money in the short term," said Josh Clay of the conservation group Big Life Foundation, which is headquartered near Amboseli. But such sales also radically alter Maasai community life. Big Life, which employs hundreds of Maasai rangers, works with communities to try to keep wildlife corridors functioning and unfenced, often paying landowners for conservation leases to keep their land open—much like the USDA is doing in Wyoming. It's working: The group recently protected a corridor called Nairrabala, to the north of Amboseli, by leasing land from a thousand Maasai owners. But the agreements are short-term, and the lure of a land sale is always there. These arrangements have also invited accusations of perpetuating colonial attitudes and relationships that marginalize pastoralist communities, in the interest of protecting wildlife.

There is also the ever-looming threat of infrastructure. One rainy morning, I set out from Nairobi on the SGR, the Chinese-built train that opened in 2018. The train crossed Nairobi National Park along elevated tracks, then rumbled for a couple of hours across farmland and communally owned grazing land, passed the occasional town, and eventually crossed through Tsavo National Park, an 8,500-square-mile expanse—about the size of New Jersey—that is actually two separate parks, Tsavo East and Tsavo West. The park is bisected by the A109 highway and two sets of train tracks. The SGR parallels older tracks, laid down by the British at the turn of the twentieth century in a project nicknamed "the lunatic express"—both for its astronomical cost and because more than one hundred Indian workers laying tracks were killed in one year by a pair of lions who became known as "the man-eaters of Tsavo."

The lunatic express, whose tracks still carry freight trains, was instrumental to the creation of Nairobi as a capital city— and even, some have argued, of Kenya as a country. As Daniel Knowles, a former Nairobi correspondent for *The Economist*, wrote in 2016, "Without the train, Kenya—a colonial confection that brought together dozens of tribes in a territory drawn with a ruler on a map—would not have come into existence."

Today, the SGR makes travel between Nairobi and Mombasa far quicker and easier than the lunatic express. But the new train has also come with a range of challenges for both wildlife and people, some of which are just beginning to become clear. As our train pulled into Voi, a rapidly developing town that serves as a gateway to Tsavo as well as an agricultural hub, we passed a neighborhood of half-dilapidated buildings on the east side of the tracks. It looked as though a bulldozer had come through and knocked some structures down and ripped the roofs off others; piles of rubble lay everywhere, and inside some demolished buildings you could see abandoned furniture and other possessions, like people had left in a hurry. The exact story differed depending on whom you spoke to, but the gist was that the entire little village of shops, houses, even a decades-old mosque, had grown into a thriving settlement on land that someone else owned. With the SGR now here, the land value had risen, and the owner had decided to cash in. The property's new owner, who planned to build a bus depot, seemed to have given little notice before arriving with heavy machinery to knock things down. Somewhere upwards of a thousand people had been displaced in a matter of days.

For wildlife, there were other problems. The SGR represents an additional barrier in a growing gamut of barriers between the two halves of the national park, part of a longer migration route stretching across southern Kenya and into Tanzania. In some places, the tracks sit atop a steep-sided berm; in others, they

are elevated on concrete bridges. Seven purposely designed wildlife underpasses cross beneath the elevated tracks, and a few dozen smaller underpasses—intended for drainage and access by SGR maintenance workers—also provide a corridor for some species. But who actually benefits from these crossing structures is uncertain. Several times a week, a team of monitors from the Kenya Wildlife Service and Save the Elephants inspects each crossing, accompanied by an armed ranger. They are looking for footprints in the dirt.

I drove beside a stretch of track with them one blazing-hot morning, to see what they were finding. Because, like in Samburu, the rainy season had been good, most animals had access to all they needed and therefore had no real reason to leave their present location. The monitoring team pointed out the footprints of baboons and mongoose—one species going east, the other west—as well as some antelope poop, as we moved from underpass to underpass, heading south. Like the Mama Tembos, these monitors upload animal-track data to an app, called Survey 123; once they've marked the tracks, they rub them out with their feet so they always know the tracks they see are new.

In virtually every underpass we checked that morning, the tracks of one species were a constant: goats. The crossings enable local pastoralists to illegally graze their livestock in the national park. It's a conundrum. Long before the parks were established, before colonization and the advent of railways and paved roads, pastoralists moved across the landscapes unimpeded. By erecting borders and fences and all the other linear monuments we've used to divide the planet into artificial

---

Two railways and a major highway bisect Kenya's Tsavo National Park. Elephants and other wildlife take advantage of this and six other purposely designed underpasses beneath the SGR, the Chinese-built train that opened in 2018. *Richard Moller/Tsavo Trust*

groupings, we've created what political ecologists have called "ecological apartheid, with stark separation enforced between mobile populations of people, domestic animals, and wildlife." It's crucial to ensure that future conservation projects don't further alienate pastoralists from their traditional landscapes. It's also crucial that the limited spaces inside parks and reserves don't become overrun with livestock that crowd out wildlife. We can't keep separating people from nature, but how to reintegrate the two sustainably may be one of the biggest challenges of the twenty-first century.

I was struck, once again, by how radically we've transformed the planet's natural systems. As I stood in an underpass beneath the SGR, the prospect of trying to conserve species and livelihoods within the existing tapestry of past decisions, both bad and good, all interwoven and lined up like dominos seemed suddenly incomprehensibly crazy.

We stopped at one of the massive wildlife underpasses designed to allow elephants and other large animals to cross. It seemed perfectly imaginable for a family of elephants to wander through here, though first they would have to walk over the lunatic express tracks. And just a few hundred meters from the underpass was the A109, filled with a constant parade of trucks going to and from the port in Mombasa. Two months earlier, an elephant had been hit by a bus and killed on the highway. Since then, one of the monitors told me, they hadn't seen any elephant tracks in any of the underpasses. It could have been because of the plentiful food and water in the park. Or it could have been that other elephants saw the tragedy and now were steering clear.

Across the highway, we could see a giant fence. It seemed to block off the landscape on that side. I asked Ewan Brennan, the head of Save the Elephants' Tsavo operations, what it was, and what the elephants did when they had successfully made it across the two rail lines and the highway. The fence, he told

me, was new. "It's just a guy fencing his land," Brennan said. "Everyone has tried to talk to him but he doesn't care. So that's it." The elephants hated walking along the highway, where truck drivers honked and yelled and harassed them. So this massive underpass built under the SGR might now be largely useless. It was just one person, one fence, one crossing point. But when the number of possible crossing points is dwindling to zero, it was huge. "On a big scale, this bit is in the middle of a mega-corridor. When you close off little choke points, then the mega-corridor becomes less viable," Brennan said.

But not everyone is building barriers. Others are trying all kinds of innovative tactics to coexist with the elephants while also maintaining their livelihood. Deep in the Sagalla Hills west of the train tracks and highway, farmer Jones Mwakima has been experimenting with a range of low-cost technology to keep elephants away from his maize, sorghum, and lentils. Mwakima said that as a kid growing up, he rarely saw elephants, and crop raiding was not an issue. The community knew exactly where the elephant corridor was, but the animals kept to that route and the people steered clear and it was mostly peaceful.

"But as the years have been going by, things have been changing," Mwakima said. "Mostly due to human activities. The elephants have been brought closer and closer to the community." It began with the construction of a big ditch, several kilometers long, for moving water, which stranded the elephants on one side. Then came the SGR. Now, he said, elephants that cross through the region to get between the two halves of Tsavo— some of the animals migrating to and from Tanzania—are more concentrated where the humans also are. "We are having more and more conflict."

Mwakima believes the elephants have a right to the landscape just as much as he does; he also sees benefits to their presence. In the years when elephants don't pass through, the land becomes

overgrown with trees and shrubs, making it difficult to graze livestock; elephants open up the landscape for grasses to grow.

He and many other farmers nearby use beehive fences to keep the elephants out of their farms; they sell the honey to STE, which in turn sells it to tourist lodges and to stores in the capital. Mwakima also employs a whole toolbox of tactics for dealing with the elephants. There are the tools for alerting people to elephants in the area—watchtowers, dogs, donkeys, strips of wire with cowbells that clang when elephants pass—and the tools to convince the elephants to go elsewhere. Mwakima uses torches, flashing them on and off to scare the elephants away. He also made a device called a "noise cannon," from old cans, nails, rubber scraps, and wood, that makes a loud grinding noise to scare away elephants. It's now part of a "coexistence toolkit" that STE hopes will help elephant-adjacent communities around the world.

The more the elephants raid crops and develop a taste for them, though, the worse the situation will be—which is why there's no substitute for keeping corridors open. "They learned that there is more nutritious food to be found. So now they have to keep on coming." Mwakima said he's seen that many of the elephants simply won't cross the SGR line. Infrastructure built without considering the impacts on either wildlife or local communities is destined to cause problems. STE and other conservation groups are trying to play a much bigger role in the planning of the new highway, using data from the underpasses to help make better decisions this time. Meanwhile, Mwakima is committed to coexisting with elephants and other wildlife, and to working with others in his community. For now, there's really no other choice.

An elephant calf mock charging a Save the Elephants vehicle in Samburu National Reserve. STE researchers can identify more than 600 elephants in the region by sight. *Courtesy of Frank af Petersens/Save the Elephants*

CHAPTER 11 *Colorado & British Columbia*

The Missing Link:
Toads, Caribou, and Us

There is a moment in the animated film *My Neighbor Totoro* where a father stands with his two daughters at the base of an enormous camphor tree. "It's been around since long ago," he tells them as they look up in awe. "Back in the time when trees and people used to be friends." I've watched the film with my family over and over, and that line gets me every time. We used to be part of nature, one with it, in friendship with much of it. (Though maybe not the man-eating lions.) We used to coexist peacefully with many other species. We wandered the planet's gorgeous landscapes together, for millennia—with elephants and elk, bear and bats, wood storks and warblers and white-lipped peccaries.

Previous Spread: Caribou from the Klinse-Za herd in British Columbia. A team of western and Indigenous conservationists has been working together to restore this herd, which lives in old-growth forests and feeds mainly on lichens. Industrial development has split caribou herds into isolated groups, and the proliferation of roads through forests makes the caribou easier prey for wolves. *Wildlife Infometrics*

But then we changed. I don't mean this in a biblical expelled-from-Eden kind of way. Eden is a myth, and the change certainly didn't happen all at once. But slowly and then increasingly quickly we began molding the planet into something different, something that benefited humans at the expense of nearly everything else. And only some humans, at that. Viewed in this light, we've been pretty shitty friends. Frenemies, at best.

Enabling those other species—the ones we evolved among, and which have just as much right to a future as we do—to stick around means ensuring their resilience. For an ecosystem, or a species, or even an individual population of animals, to be resilient, it needs the ability to bounce back from stress, disturbance, or catastrophe. By its very nature, resilience requires contingencies. There has to be more than one path for an elephant. Because if there is only one possible route, then a single person's whim—the choice to build a fence, or a farm, or a border wall—can doom an entire population of animals. If that is where we are headed on a global scale, we are in serious trouble. It's time for us, as a species, to be allies with nature again. We must find ways to expand, rather than continue to limit, the available options for other species. The more we reduce those options, the more work we create for ourselves.

Throughout many years of reporting, I've seen the same pattern appear: As we cut off the planet's connections, leaving fragmented landscapes and broken systems, we've left ourselves to complete the circuits and fill in as the missing link—by any means necessary. The means might involve gathering community members under a baobab tree to ask for a thirty-meter elephant pathway, or planting tens of thousands of trees, or taking down unnecessary fences, or lovingly sprouting and tending agave seedlings. Or they might involve engineering complicated, expensive, labor-intensive, farfetched solutions just to accomplish what nature once did, you know, naturally.

— - - —

Ten thousand feet above sea level one early summer afternoon, I donned waders and gloves and walked into a shallow lake in search of toads. They were surprisingly easy to catch—you just reached out and scooped them up. The researchers I'd accompanied to this northern Colorado lake showed me how to zip the placid toads into little mesh bags that we carried to shore, keeping the animals captive just long enough to identify them, weigh them, and swab their skin before releasing them back to their alpine wetland home. The toads seemed to know the drill; they'd done it before. They might have even recognized the scientists, like the scientists recognized them. The toads had tiny microchips embedded just beneath their skin, but the team working to ensure the survival of *Anaxyrus boreas boreas*, the boreal toad, could identify most of the animals by the precise pattern of spots on their soft bellies.

Like many amphibians around the world, boreal toads have been hit hard by a type of fungus called *Batrachochytrium dendrobatidis*, or Bd—also known as a chytrid fungus—which causes a deadly skin disease. Bd has infected more than 350 species of amphibian; it has caused extinction or disastrous declines in at least 200 types of frogs. Scientists have called the carnage "the greatest disease-caused loss of biodiversity in recorded history."

Native to the Rocky Mountains from Utah up to Canada, boreal toads began to decline in the 1970s. By the early 2000s, scientists had identified Bd as the cause. The toads in Colorado have fared the worst, though no one is quite sure why. Bd has afflicted the state's boreal toads at one site after another. Of more than a hundred previously documented breeding sites in

A biologist wears protective gloves while holding a boreal toad to prevent the transmission of a fungus called Bd, also known as a chytrid fungus. *Glacier National Park Service*

Colorado, where toads from different home sites gather to mate, only a fraction are still active.

"We have to assume that eventually it's gonna be everywhere," said Harry Crockett, the Native Aquatic Species Coordinator for the Colorado Department of Parks and Wildlife, about the fungus. "So, we are ultimately going to need animals that are tolerant or resistant to it." Scientists are cautiously hopeful at least some amphibians can evolve that resistance; some individuals appear to recover from an infection, even as their compatriots succumb. But in the meantime, if too many toads perish, or if whole populations disappear, the species could lose vital genetic diversity, leaving it vulnerable to another disease or catastrophe down the line. Crockett oversees a project aimed at avoiding such a crash, trying to bring resilience to the toads by ensuring large enough chytrid-free populations to maintain a diverse gene pool.

"We're buying enough time for it to evolve some tolerance or resistance," he said. "What we're trying to do is establish more populations, so we outpace the rate at which they're being lost." Research by an evolutionary ecologist at Colorado State University named Chris Funk showed that genetic diversity among Colorado's remaining toad populations was still high; it's a sign of interbreeding populations from before the species went into free fall, "a signature of genetic connectivity across the landscape."

When biologists first learned that Bd was behind the toad declines, they collected toads from many of the remaining breeding sites and brought them to a state facility dedicated to

Life may seem promising for these boreal toad tadpoles, but like many amphibians around the world, the toads have been hit hard by a fungal infection called Bd, which causes a deadly skin disease. Researchers in Colorado, where the toads have fared the worst, are trying to maintain fungus-free populations with diverse gene pools. *Shane Gross/NPL/Minden Pictures*

restoring threatened species. There, they could breed toads in captivity and still maintain the needed diversity. It also meant they could use so-called nearest neighbor sources—toads from close-by areas—to start populations in new habitat. Colorado's boreal toads are listed as threatened under state law, and part of the recovery plan involved relocating Bd-free toads to places that seemed like excellent toad habitat but didn't already have any toads—or the fungus.

At the northern Colorado wetland I visited, near the top of a mountain pass west of Fort Collins, the state had stocked thousands of tadpoles beginning in 2008. In 2014 they first documented breeding in the area (male boreal toads reproduce at age four, females at six). And in 2016 they found tadpoles in the lake. The new population was doing well so far, and the fungus hadn't yet appeared. Researchers regularly returned to keep tabs on the toads, check for chytrid, and tag any new individuals they found.

Moving species around has become an increasingly common conservation strategy. Scientists have used the tactic to help Canadian pine forests adapt to climate change by introducing seeds from more southerly parts of the trees' range that could help northern trees better tolerate heat and drought. They've relocated panthers from Texas to Florida, to inject vital genetic diversity into an isolated, inbred, and rapidly dwindling panther population. They've moved Columbia spotted frogs in Nevada to secluded mountain lakes to keep them safe from invading hordes of bullfrogs. In a world of lost connections, humans are stepping in to serve as a necessary bridge.

"When I was a grad student, I was a purist," Funk told me. "I thought it was important to understand the distinctive populations and keep them separate. Now I see we don't have the luxury of time anymore. We need to do more management, moving species around."

— – – —

A few years ago, for a magazine story, I witnessed a particularly extreme variation on this idea. Several hours up a steep and narrow active logging road near the tiny town of Hudson's Hope, British Columbia, I arrived at a flat clearing where a high fence wrapped in black fabric marked the edge of a thirty-seven-acre patch of woods and meadows. There, a dozen female woodland caribou and their nine newborn calves were spending the summer. This "caribou maternity pen" was the centerpiece of an effort to save a declining herd, on behalf of two Indigenous communities that had traditionally relied on caribou for food, clothing, medicine, and tools. Since childhood, Roland Willson, the chief of the West Moberly First Nations, had been listening to tribal elders reminisce about hunting caribou on their traditional lands, something he'd never had the opportunity to witness or take part in himself.

About half of Canada's roughly sixty herds of woodland caribou were shrinking, and many were in "rapid decline," decreasing by half every eight years. A type of deer, caribou live across a huge slice of the planet's North, from the Arctic tundra south through the boreal, or northern, forest. Woodland caribou, a subspecies, are in the direst straits. They live in old-growth forests, feeding largely on lichens. The boreal forest is still the largest unbroken forest on Earth, representing a quarter of all remaining intact forest. But nearly a third of Canada's boreal has already been carved up or earmarked for industrial use—logging, mining, oil and gas extraction.

Cutting down the forest wipes out caribou habitat and splits herds into smaller, disconnected, unsustainable groups. Building roads across forests provides easy access for wolves, which can travel up to three times faster along roads than through unbroken forest. Because caribou reproduce slowly—females are pregnant for nearly eight months and give birth to just one baby

at a time—the problem boils down to simple math. Too many caribou are dying, and not enough are surviving to reproduce.

The maternity pens are designed to change the math, essentially serving as a kind of connective tissue suspended in time. Each spring, a team of people locates the caribou herd and then drops nets from a helicopter to capture the pregnant females. Scientists on the ground sedate the animals, zip them into body bags, and load them into the helicopter, which flies them to the fenced enclosure, where they will give birth. The mother caribou can raise their babies safe from prowling wolves and grizzly bears, guarded by First Nations members who live in a cabin on the premises and walk the perimeter several times a day, looking for any compromises to the fencing and keeping watch for predators. In late summer, when the babies are old enough to outrun a grizzly, if not necessarily a wolf, the caribou are released back to the wild.

When the forest was intact, the caribou herd would migrate high in the mountains during the winter, where snow would act as a buffer to wolves. Opening the forest provided wolves a corridor straight to the caribou's winter range, and also created extra habitat for deer, elk, and moose. With more to eat, the wolves thrived, increasing the threat to caribou. Changes to the connectivity of the landscape had created clear winners and losers. Humans were now attempting to serve as a temporal link for the caribou, keeping them safe until enough time had passed that the calves might survive. The maternal pen was Noah's Ark in the boreal forest, Atlas holding the caribou world aloft.

A pregnant female caribou is sedated and loaded into a helicopter to be transported to a specially constructed "maternity pen," where she can give birth and raise her young in safety. The extremely heavy-handed intervention has succeeded in bringing herd numbers up. But protecting caribou habitat from industrial development seems like a far more straightforward solution. *Landon Birch/Wildlife Infometrics*

The Canadian scientists and First Nations partners I spoke with all believed that the intervention, though preposterous, was necessary to keep the herd from collapsing. "Because of where we are, we have to be a little heavy-handed until we can get things back into balance," Willson told me. Most importantly, the project (which also included culling a large number of wolves) was working. The caribou herd was growing again, up from a low of just 16 animals in 2013 to 140 in the summer of 2022, when I checked back with Scott McNay, the Canadian biologist who'd taken me to see the maternity pen. But it seemed so extreme, such a huge undertaking—when simply protecting the caribou's habitat would have been far more simple. Humans had designed a clever and ingenious workaround to the problems caused by a highly fragmented landscape, but they could have instead worked harder to keep it intact.

McNay agreed wholeheartedly; he felt strongly that the maternal pen was only an emergency solution. "We're not going to continue these things forever," he said. Hopefully he wouldn't have to do it much longer: Since we'd last spoken a few years earlier, there had been a big expansion in the amount of protected area in the region, thanks to a new agreement between the province and the First Nations. Under Canada's Species at Risk Act, akin to the US's Endangered Species Act, if provinces aren't fulfilling their conservation obligations, the federal government can step in. Frustrated by British Columbia's lack of action to save caribou, which are protected under the federal law, Prime Minister Justin Trudeau's government declared an "imminent threat" to the province's herds. If BC didn't act, the federal government could take control of the province's natural resources.

Caribou enjoying a safe summer season in a maternity pen in central British Columbia. Caribou are listed as endangered under Canada's Species at Risk Act, which has provided an opening for rewilding and reforestation efforts. *Katie Orlinsky*

The new land protections gave McNay and his team a rare opportunity to restore a former industrial landscape. The team was turning old logging roads, snowmobile trails, and seismic lines from oil exploration back into forest, trying to knit the whole system back together. Meanwhile, First Nations had been taking the provincial government to court over illegal development on their lands, and winning. "The most interesting thing about this whole project is not the biology," McNay said. "It's having the province take responsibility for their obligations" to First Nations, through existing treaty agreements.

Rewilding roads, though, came with its own challenges. McNay's team had closed some roads to winter recreation, but snowmobilers had simply ignored the closures. It was difficult to get permits from the province for the restoration work, which didn't seem to fit any of the existing operational categories, and it was also hard to find workers to hire. "There are literally tens of thousands of kilometers of road that are available for restoration," McNay said, "and nobody has experience doing this work." His goal was to restore fifty to a hundred kilometers (about thirty to sixty miles) a year—digging up the compacted roads, loosening the soil, planting trees, bending or felling some existing trees to cover up or block a road, since it would be a decade or more before the new trees made the pathway truly impassable. "Sometimes we might take extra effort at the start of a road to make it extra inaccessible."

McNay was also working to restore a larger caribou herd that the small herd had once been part of, before the groups became isolated from one another. "We've gotta get our minds out of 'herd by herd by herd' recovery, and think about how do we increase the metapopulation," he said. "It's exciting, having opportunities to change an industrial landscape into something more natural."

Despite the challenges, despite the undeniable absurdity of the project he'd undertaken for more than a decade, despite

setbacks and lawsuits and hate mail, McNay sounded optimistic. He'd long ago decided he was done with "planning and research and collecting more data" while more forest was destroyed and more caribou herds declined. Now he was on the ground doing things. Action is invigorating. That was another message I heard everywhere I traveled. The most optimistic people were the ones working with specific communities to protect specific animals in specific places.

Conservation, on some level, will always be about holding the line, playing defense as attacks continuously materialize. But as global warming alters people's lives around the planet, we have a unique opportunity to change our relationship to nature. If we continue to act as though humans are the only species on the planet, it will become something of a self-fulfilling prophecy. We can no longer act selfishly. We need to recognize that when we reshape Earth to suit our own needs, it impacts, directly and indirectly, the rest of the world's creatures—which also have a right to live, breathe, eat, love, roam, and persevere. You could question the morality of trying to save animals when wars are being waged and people die of starvation and preventable disease every day around the world. But you could also argue that it's all related, and that we must embrace a mindset of care and compassion—for one another as well as for other species—and learn to think more expansively about the impact of all of our actions.

In a time of increasing fragmentation and declining connectivity, we need to recognize that one person's fence can doom a whole species—and also that one person's fence removal can reopen a world of potential. We each may hold more power than we know, to function as essential links in an otherwise broken chain.

The Power of
Bearing Witness

In Iceland, two decades ago, before my husband, Phil, and I were married, we backpacked with a group of environmental activists on a pilgrimage through a dreamscape of meadows and wetlands that would soon become a lakebed. The massive and controversial industrial project underway included three reservoirs and five dams—one of which, Kárahnjúkar, would be the largest in all of Europe. It was being built to generate electricity for what would be Iceland's largest power plant. In total, the project would cover nearly 400 square miles—all to power one aluminum smelter, run by the American corporation Alcoa. It was, and still is, difficult to fathom. Icelanders had fought the project for years, to no avail.

Previous Spread: The thundering Jökulsá í Fljótsdal, a glacial river in Iceland's eastern highlands. It was dammed, and twenty-two square miles of highlands flooded, to build the Kárahnjúkar hydropower plant, which runs a single aluminum smelter. *Christopher Lund*

I remember clearly how, late one Arctic summer night, as I'd brushed my teeth in the barely dimming light, I wrestled with whether it was okay to spit my toothpaste out on the tundra grasses. After all, they would soon be at the bottom of a reservoir, so what difference did it make? Still, it seemed somehow wrong, disrespectful.

What do you do, though, in a place whose days are numbered? For our two guides and the thirty-odd Icelanders who had joined them on that journey, the answer was to bear witness. It's a message I've returned to again and again in my years of reporting on environmental change. Sometimes it's all that we can do. So the group had walked, twelve miles a day, up and across Iceland's remote eastern highlands.

We'd started our highland pilgrimage on the east side of the Jökulsá í Fljótsdal, a glacial river, where the jeep tracks ended at one of Iceland's many abandoned farms. The river's eastern banks were coated in lush grass and sheep-sized tussocks broken at regular intervals by streams that tumbled as waterfalls down the steep hillside. Critics of the dam had said the scores of similar waterfalls on the western side would dry up once the power plant, located inside the western bank's sloping mountain, went online.

One day as we paused for a snack, I took in the view all around. To the east, the golden face of Snæfell, the 6,000-foot glacier-topped volcano, shimmered in the mid-morning sun, its summit still partially hidden by a cape of fog. To the south, rivers of ice spilled from the edge of Vatnajökull, a hulking glacier so large it creates its own weather system. To the west stretched wetlands, miles of squishy, mossy ground littered with reindeer antlers and goose droppings and the discarded eggshells of baby birds. In the small lakes that were everywhere swam a tiny, ray-like creature that etches sand tunnels with its tail and once coexisted with dinosaurs.

Little land birds flitted through the grass, and on the drier patches of ground grew tiny white tundra flowers, a favorite snack of the pink-footed goose. Smooth white mushrooms also grew, a kind whose Icelandic name means "cheese of the field." Skeptical at first, I reluctantly bit into one my Icelandic companions passed around, and it did, in fact, taste like cheese. The water I drank to wash it down had come straight out of a stream we'd passed earlier that morning. We were walking, barefoot, across this landscape unbroken by a trail, when the peace was interrupted by the sound of a dynamite blast.

It came from the north: Somewhere just out of view in the direction we were headed, dump trucks and heavy machinery were erecting the 633-foot-high dam. "I hope that's the elves stopping the project," someone in the group said, and they all agreed. As we trekked closer to the canyon edge, the "beep beep beep" of trucks going backward began to mingle with the "peep peep" of the golden plover, sounding its alarm call in the heather. That night at camp, one of the guides, Ásta Arnardóttir, who taught yoga in Reykjavík, led the group in a short session of stretching, swinging, and whooping. She taught a pose she called "the reindeer," in honor of the wild herd that sometimes grazed this area but was then across the Jökla in another wetland called Kringilsárrani. Scientists predicted that the herd, which would be penned in on one side of the lake, would shrink.

As we stood with one leg partially bent, the other knee raised in front in a move reminiscent of *The Karate Kid*, our hands making antlers above our heads, Ásta spoke—as she often did—about "augnablik." It was the name of the trekking company she

---

Protesters who had trekked for several days across the Icelandic highlands held hands—and many cried—as they witnessed construction of the Kárahnjúkar Dam in July 2005. The project was completed and the reservoir filled four years later. *Hillary Rosner*

and Ósk Vilhjálmsdóttir, an artist and our other guide, had created to take people to the highlands so they could see for themselves what was being lost. Augnablik translates as "moment" or "blink of the eye." "When we come into nature we fall into the moment," Ásta said. "We realize how small we are, that life is short, that the moment is important." It's another facet of umbilical connectivity.

Shortly before 4:00 a.m. the next day, awakened by the slight brightening that follows the brief dusk of an Arctic summer, I emerged from my tent. I could see glimmering lights about an hour's walk away: bulldozers, already working. Then another blast disturbed the dawn.

Later we reached the viewing platform above the construction pit, where tunnel-borers sliced through rock and an army of dump trucks carried shattered bits of canyon wall down a series of switchbacks. More than fifty protesters stood and linked arms, and many of them cried. I joined the circle. The man to my right, a longtime Kárahnjúkar activist who had never before seen the site, sobbed loudly. Iceland had a population of 280,000 at the time, most of them connected by only a degree or two of separation. These protesters felt an acute sense of betrayal. A middle-aged woman on the trek said it was "like being betrayed by kin."

We left the platform, and the blue bus that had been meeting us each evening with our gear drove us over a shiny new bridge, depositing us on the banks of the Kvislar Sauda, a smaller glacial river that tumbled through gentle waterfalls and collected in enticing flat-bottomed pools before meeting the raging Jökla. The sky was a radiant blue as we walked east along the Sauda in a valley that looked like a scene out of Tolkien. Bright green moss carpeted the hillside, and fuchsia flowers grew in sunny patios ringed by rocks. We lay down and dipped our faces into the river and drank it straight out of our hands. I understood

that I was likely among the last hundred or so people who would ever walk along that river. Everything within view would soon be submerged.

I couldn't change the fate of that landscape and the creatures that relied on it for survival. Nor could the protesters. But we could bear witness to its beauty and then to its demise. We could mark the forces of greed and apathy that enabled its obliteration. The Icelanders I walked among knew it was too late to stop the dam, but they walked anyway. The trip changed them, and they shared what they saw, and they found new connections to the environment around them. In those umbilical connections lie power and possibility.

We can all do this. We can go outside and watch what is happening in—and to—the world around us. We can tell others, and explain why it matters. We can take action in our own communities, or find projects or places or species to champion. We can all bear witness. And most important, we can care.

Next Spread: A black bear navigates human infrastructure—a barbed-wire fence—in the Everglades as it follows a game trail between Big Cypress National Preserve and a privately owned protected area. *Carlton Ward Jr.*

ACKNOWLEDGMENTS

I worked on this book for far longer than I initially intended, and I had help from a great many extraordinary people along the way. It makes me smile just to think about all of them.

Thank you to the team at Patagonia Books: Sharon AvRutick, Jane Sievert, Kyra Kennedy, Christina Speed, Sonia Moore, Michele Bianchi, John Dutton, and the amazing Karla Olson, who believed in the project from the start. My eternal gratitude to Makenna Goodman, who set the whole thing in motion and so much more.

This book draws on reporting I've done throughout the years, some of which was enabled and supported by brilliant editors. I'm grateful to Tim DeChant, Michelle Nijhuis, Paul Tullis, Rob Kunzig, Steven Bedard, Alan Burdick, Elizabeth Royte, and Katie Courage for editing and publishing my connectivity-related stories.

I wrote parts of this book on much-needed retreats with inspiring writers. Thanks to Erin Espelie, my steadfast writing buddy, and to everyone else who wrote with me: Florence Williams, Stephen Miller, Peter Brannen, Moe Clark, Laura Krantz, Scott Carney, Steven Bedard, and Robert Karjel. Thank you to Jennifer Leitzes and Jon Hoeber, Hannah Nordhaus, Michael and Maria Ainbinder, and the staff at Osa Conservation for providing wonderful writing spaces. I am also deeply indebted to Hannah Nordhaus and Florence Williams for editing early chapter drafts, and for frequent book advice and unwavering moral support.

I'm grateful to all the dedicated scientists, conservationists, government officials, and others who shared their time and ideas with me, hosted me, patiently answered my endless

questions, and generously allowed me to traipse around with them. Extra special thanks to Josh Tewksbury, Dave Theobald, Nick Haddad, Andy Gonzalez, Andy Whitworth, Eleanor Flatt, Gabriela Vinueza, Angela Tavone, Francesca Cagnacci, Arthur Middleton, Nick Lapham, Jane Wynyard and the STE crew, and Amanda Paulson, who was a constant sounding board. Thanks also to Bernhard Warner and family for hospitality in Rome, and to Shermin DaSilva, whose fieldwork I sadly had to miss at the eleventh hour.

The idea for this book took shape during a Ted Scripps Fellowship. For that wonderful year, I thank the Center for Environmental Journalism and Cindy Scripps. I benefited from research help early on from Shannon Mullane, and later from Devin Farmiloe, who is also a spectacular travel companion when bats fly out of your toilet or rising rivers necessitate frenzied predawn escapes.

Finally, none of this would have been possible without my family, who always look out for me: My husband took on extra parenting, pet care, and household duties during my many reporting trips and provided me with the world's best coffee, cocktails, and meals when I returned. My son let me yammer on incessantly about wildlife connectivity and conservation, and his curiosity about the world is a constant source of hope. My sister always makes me feel like a far braver adventurer than I am. Thanks to my father, a consummate autodidact, for inspiring me to learn about the natural world and always championing my writing. And to my mother, who would not have set foot in a rainforest if you paid her but would have been this book's most vocal publicist: I wish you were here to read it.

RESOURCES

"I hope this book sparks your own curiosity, and inspires you to help protect our natural world, in any way you can."
—from the Author's Note

Scan this QR code for more information about the critical work being done by the researchers and organizations mentioned in this book, as well as other related material:

- Maps
- Websites
- Author's endnotes
- Reader's guide
- eBook
- Audiobook

Previous Spread: An adult peregrine falcon stands guard over chicks nesting in a window box on a Chicago balcony. Wildlife will use resources wherever they can find them. Since peregrines mainly hunt other birds, a city full of pigeons makes for easy meals. *Luke Massey*

INDEX

A

Absaroka Fence Initiative (AFI), 187, 189
Absarokas, 178, 187
Adamello-Brenta Nature Park, 163
Adige Valley, 168, 170
agaves, 119–39
Alagona, Peter, 51
Albrecht, Glenn, 45
Alcoa, 334
Alley Pond Park, 211, 215, 219
alligators, 250–51, 262, 277, 282, 284
Alps, 161, 163, 165, 166, 168
Amazon Biodiversity Center, 38–39
Amazon rainforest, 37–39
Amboseli National Park, 307, 308
American Bird Conservancy, 73
American coots, 282
AmistOsa, 92
anhingas, 277
antbirds, 38–39
anteaters, 95, 111
anthropogenic resistance, 148–49, 160, 165, 304
Apennine chamois, 148, 149, 160
Apennines, 2, 142, 143, 144, 146–49, 158
apple snails, 279
arboreal bridges, 114–15
Arctic terns, 35
Ásta Arnardóttir, 336
Audubon Society, 74, 243, 244, 246, 251, 253–54, 262, 273, 275

B

badgers, 199
Bai, Jin, 234, 236, 238, 239, 241, 244, 246
bandicoots, 183
Bat Conservation International (BCI), 118, 120, 124, 128, 129, 137
Batrachochytrium dendrobatidis (Bd), 318, 323–24

bats
    lesser long-nosed, 128, 131
    Mexican long-nosed, 118, 123–24, 127–29, 131
    population decline of, 128–29
bearded dragons, 183
bears
    black, 30, 210, 339
    brown, 146, 161, 163, 165–68
    grizzly, 44, 144, 146, 191, 194
    Marsican, 2, 142, 143, 144, 146, 149, 152–58, 160
Beirne, Chris, 92, 113, 148
Belt and Road Initiative, 66, 293
Benavides, Rodrigo, 104
bettongs, 183
Biden, Joe, 264
Big Cypress National Preserve, 268, 282, 339
Big Life Foundation, 308
biodiversity (biological diversity)
    agriculture and, 55
    history of, 37
    social inequality and, 224–30, 232–36, 238–39, 241
biogeochemical cycles, 187
Bipartisan Infrastructure Law, 6, 67
birds. *See also individual birds*
    connectivity and, 71–74
    counting, 242–44, 246
    migratory, 71–74, 76
    tracking, 74, 76–77
bison. *See* buffalo
black-bellied plovers, 74, 76
Black Lives Matter, 226
blue morpho butterflies, 98
bobcats, 30, 71, 210
Bonnie, Robert, 191, 194, 195
boreal toads, 318, 323–24
Brennan, Ewan, 312–13
Bronx Park, 213

Bronx Zoo, 11
Brudvig, Lars, 60–61
Budd, Bob, 195
buffalo (bison), 30, 146, 176, 177, 180, 184
Burmese pythons, 281–82
butterflies
blue morpho, 98
monarch, 35, 59, 118, 246

C

Cagnacci, Francesca, 161, 165–70, 172
Caloosahatchee River, 255
Campos, Alberto, 111
Caragiulo, Anthony, 205–6, 208–11, 219, 220
caribou, 318, 325–26, 329–30
Carolina willow, 280–81
Caspi, Tal, 247
catfish, 277, 279
Center for Large Landscape Conservation, 63
Central Park, 10, 11, 14, 22, 198, 206, 218
Chalío, Don, 118
Chávez, Hugo, 100
Cheney, Robert, 57
Chicago, 203, 205, 216
Chimney Hollow Reservoir, 27–28, 31, 33, 44–45
chimpanzees, 291
Christmas Bird Count, 243
chytrid fungus. See *Batrachochytrium dendrobatidis* (Bd)
Cipollone, Mario, 152–53, 155, 160
Clay, Josh, 308
Clem, Shawn, 250–51, 253–55, 268–70, 276–77, 279, 281–82, 284–85
climate change
corridors and, 84, 91, 95–97, 115
effects of, 12, 16, 28, 96–97, 262
future movement and, 35
importance of, 13
climate resilience, 97
Clinton, Bill, 258
coatis, 83, 203
Colorado–Big Thompson Project, 27, 31
Colorado River, 12, 27
CONANP (Comisión Nacional de Áreas

Naturales Protegidas), 121, 129
connectivity
as circulatory system of nature, 63–64
complexity of, 23, 254–55
corridors and, 49, 63
definition of, 18
ecological, 19
emotional and cognitive, 19
human behaviors and, 148–49
importance of, 34–35, 45, 77
landscape permeability and, 78–79
metapopulations and, 36
research on, 38–39, 42
umbilical, 18, 338, 339
conservation
bearing witness and, 339
community-based, 107, 115, 120
importance of, 7, 176
public support for, 273, 275
Conservation International (CI), 94
Corcovado National Park, 89–92, 94–98, 104–5, 190
Cordischi, Fabrizio, 147, 157
Corkscrew Swamp Sanctuary, 251, 253–55, 261, 262, 265, 268–70, 273, 275–77, 280, 284–85
Cornell, Bradley, 273, 275–77
Corradini, Andrea, 161, 165–68
corridors
climate change and, 84, 91, 95–97, 115
connectivity and, 49, 63
definition of, 48–49
effectiveness of, 50, 54–56, 59–60
matrix and, 59, 77
officially designated, 91–92
planning, 77–79, 91–92, 108–10, 113–14
plants and, 60–61, 63
cougars. See mountain lions
coyotes, 10, 14, 21, 22, 198–221, 247
Crockett, Harry, 323
Cueva del Diablo, 124
Cumulative Outdoor Activity Index, 165
cypress trees, 251, 253–54, 282

D

dams, 17, 27. See also *individual dams*
Damschen, Ellen, 60–61

Danum Valley, 42
Darién Gap, 202
deer. *See also* caribou; reindeer
    migration patterns of, 170
    mule, 180, 195
    red, 144, 147, 148, 169
    roe, 169–70, 172
    white-tailed, 264, 281–82
    wildlife-friendly fencing and, 187, 189
defaunation, 110–11
deforestation, 13, 39, 64, 92, 98, 109, 203
Denali National Park, 42
DeSantis, Ron, 261
DeSousa, Rodrigo, 100–101, 103–4
dingoes, 183, 184
dispersal, 35, 216
Dog Fence, 183
Dolomites, 147
Douglas, Marjory Stoneman, 250, 251, 255, 258, 259
Douglas-Hamilton, Iain, 291
Dwingelderveld National Park, 67

E
eagles
    crested, 110
    harpy, 110, 111
Earth First!, 143
East River, 206, 213, 214
eBird, 243–44
egrets, 262, 265, 282
elephants
    adaptiveness of, 300
    causes of deaths for, 291, 292
    coexistence of humans and, 292, 297, 300, 304, 306–8, 313, 315
    drought and, 289
    fences and corridors for, 70, 302–4, 306–8, 310, 312–13
    gullies and, 302–4
    migration of, 288, 291–92, 294, 296–97, 300, 309, 353
    poaching and, 291, 292
    tracking, 289, 291
elk, 2, 33–34, 191, 194
Ellis Soto, Diego, 242–44, 246
Environmental Defense Fund, 194

environmental justice, 227, 229
Estenoz, Shannon, 264
evapotranspiration, 269
Everglades
    challenges faced by, 7, 262, 264
    connectivity and, 284–85
    draining of, 253, 258, 261, 279
    fencing in, 339
    restoration of, 7, 258–59, 261–62, 264–65, 272, 279, 285
    as "river of grass," 250, 251, 255
    size of, 255
    water levels in, 265, 268–70
Everglades Coalition, 259, 265
Everglades National Park, 261, 281
Everglades snail kite, 279
evolutionary rescue, 51
Ewaso Ng'iro (Nyiro) River, 288, 289, 297
extinction debt, 36–37, 242

F
Fair Housing Act, 233
fencing
    history of, 177
    impact of, 176–78, 180–81, 183–84, 186
    invisibility of, 178, 180–81
    prevalence of, 176–78, 181
    removal or modification of, 154, 178, 186–87, 189–90, 195
    wildlife-friendly, 187, 189
Ferry Point Park, 206, 213
Flatt, Eleanor, 114
Florida Panther National Wildlife Refuge, 275
Florida panthers, 251, 272, 275, 281, 284
food webs, 87, 89
Foreman, Dave, 143
Forsyth, Adrian, 94–97
foxes
    bat-eared, 183
    crab-eating, 202
    red, 183
Friends of the Osa, 95. *See also* Osa Conservation
Front Range, 21, 26, 27, 28
Funk, Chris, 323, 324

G
Galván, María Isabel García, 131–32
gentrification, 239
Ghoddousi, Arash, 149
giraffes, 66, 183, 184, 294
Glidden, Joseph, 177
Global Biodiversity Information Facility, 244
Global Heatmap, 165
Global Human Footprint Index, 180
Global Initiative on Ungulate Migration, 172
Golden Gate Estates, 270, 272, 275–76
Golfo Dulce, 82, 84, 90, 107
Gonzalez, Andrew, 36, 37, 49–52, 54, 169
Goodall, Jane, 291
gorillas, 12
Gotham Coyote Project (GCP), 205, 209–11, 213, 216, 220
Grand Teton National Park, 2, 199
Gran Sasso, 143
Great Acceleration, 14, 16
Great Basin, 13
Greater Yellowstone Coalition, 189
Greater Yellowstone Ecosystem (GYE), 187, 189, 191, 195
Green, Rhett, 253
Grunwald, Michael, 270, 272
guanacos, 184, 186
Gulf American, 270

H
habitats
    expansion of, 199
    fragmentation of, 31, 36, 37, 39, 42, 50, 52, 56–57, 121, 319
    leasing, 191, 194
Haddad, Nick, 49, 55–57, 59–60
Hanski, Ilkka, 36
Harrison, Autumn-Lynn, 74, 76
Henger, Carol, 209
Henry Hudson Bridge, 10
herons, 230, 262, 265, 279, 282
Home Owners' Loan Corporation (HOLC), 227, 228, 232
Homestead Act of 1862, 177
Horsepen Strand, 275, 285

Hudson, Henry, 52
Hudsonian godwits, 71

I
Ibarra, Ana, 118, 120–21, 124, 128, 129, 134, 137, 138
ibises, 262, 265
International Energy Agency, 64
International Union for Conservation of Nature (IUCN), 19, 163
island biogeography, 35–36, 37, 50

J
Jackson, Andrew, 264
jaguars, 89, 90, 149, 186, 200, 202
Jökulsá í Fljótsdal, 334, 335

K
Kaanki, Samuel Ole, 307–8
Kárahnjúkar Dam, 334–36, 338–39
Katti, Madhusudan, 225, 234–36, 239, 241, 243–44
Kays, Roland, 200, 202–3, 219
Kenya Vision 2030, 292–93
Kibaki, Mwai, 292
KiliAvo, 307
King, Lucy, 70, 300
Kissimmee River, 258, 261, 265
Knowles, Daniel, 309
Kremen, Claire, 55
Kvislar Sauda, 338

L
La Amistad, 91–92, 97
Lago di Scanno, 158
LaGuardia Airport, 218, 219, 220
Lake Nakuru National Park, 181, 183
landscape permeability, 78–79
land-use change, 13–14, 16–18
Large Carnivore Initiative for Europe, 163
Laurance, William F., 64–65, 67
League of Conservation Voters, 261
Lear, Kristen, 120–21, 128, 129, 131, 134, 137, 138
leopards, 66, 149
Levey, Douglas, 59
Levins, Richard, 36
Leyequien, Lissette, 121, 123, 131, 133–34

limpkins, 279
Loboitong'o, Mpayon, 304
Locke, Dexter, 233
Locke, Harvey, 44
Loloju, Benjamin, 293–94, 300, 302–4
Long Island, 209–10, 213
longleaf pines, 57, 59, 61
Loole, Jerenica, 306–7
Los Angeles, 71, 226, 228, 230, 232–33, 235, 238
Los Angeles River, 230
Lovejoy, Thomas, 37–39, 42
Loxahatchee River, 253
luxury effect, 225–26, 235, 241

M
Maasai Mara, 184
MacArthur, Robert, 35–36, 37
Majella National Park, 149
Mama Tembos, 304, 306–7
Manu National Park, 87
Mapping Inequality, 227
margays, 90
Martínez, José Jesús Reyna, 134–35
Martínez, Martina Pérez, 132–33
maternity pens, 325–26, 329
matrix
    definition of, 59
    importance of, 77, 79, 91
Max Planck Institute, 76
McNay, Scott, 329–31
Mehrabi, Zia, 77
Mendoza, Aurelio Gaytán, 137–38
Mendoza, José Inocencio Moreno, 133
metacommunities, 37
metapopulation theory, 36–37
Mianus River Gorge, 211, 213
Miccosukee Tribe, 264–65
Middleton, Arthur, 187, 189–91, 195
Mireles, Bertha Alicia Estrada, 134
misanthropes vs. synanthropes, 203, 205
monarch butterflies, 35, 59, 118, 246
Mong, Tony, 186–87
Montreal, 50–52, 54
Moore, Gordon, 95, 96
moose, 51, 190, 194, 195, 199, 326
mosquitos, 246

mountain goats, 199
mountain lions (cougars), 31, 71, 199, 200, 205. *See also* Florida panthers
Mount Kenya National Park, 70
movement ecology, 113
Mule Deer Foundation, 189
Musk, Elon, 119
Mwakima, Jones, 313, 315
My Neighbor Totoro (film), 318

N
Nagy, Chris, 210, 211, 213–16, 218–19
Nairobi National Park, 66, 308
Naja, Melodie, 262
National Audubon Society. *See* Audubon Society
National Elk Refuge, 2
nature
    changing relationship to, 331
    connection to, 23, 318
    connectivity as circulatory system of, 63–64
The Nature Conservancy (TNC), 57, 189, 190
New York City, 10–11, 14, 52, 198–99, 205–6, 208–11, 213–16, 218–19
Ngare Ndare Forest Reserve, 70
Northern Water, 26–28
Noss, Reed, 144

O
ocelots, 114, 186
Odum, Eugene, 56
Okavango Delta, 184
Okeechobee, Lake, 258, 265
Oostvaardersplassen (OVP), 144
Osa Conservation, 84, 86, 90, 92, 94–96, 98, 100–101, 103–4, 107, 113
Osa Peninsula, 82–84, 89–91, 94, 104, 107, 113, 353
Osceola, William J., 265
Ósk Vilhjálmsdóttir, 338
ovenbirds, 52, 54

P
Panama Canal, 202
Pan Borneo Highway, 66–67
Parco Nazionale d'Abruzzo, Lazio e Molise, 146, 153

Pathways to Science, 246
Paulson, Amanda, 38
peccaries
    collared, 90
    white-lipped, 89–90, 104–5, 107, 111
Pelham Bay Park, 213
peregrine falcons, 346
Picayune Strand, 272
Piedras Blancas, 90, 111
pronghorn antelope, 48, 172, 180, 187, 194, 199
Pyrenees, 42

Q
Queens Zoo, 10, 14

R
raccoons, 199, 215, 219, 279–82, 284
racism, systemic, 224–30, 232–33, 246–47
Ramirez, Manuel, 94–96
Randalls Island, 214
razorback suckers, 12–13
Reale, Valerio, 153
redlining, 227–29, 232–34, 236, 238, 241, 242
reforestation, 98, 100–101, 103–4
Reid, John W., 38
reindeer, 335, 336
rewilding, 101, 111, 143–44, 146
Rewilding Apennines, 147, 152–58, 160
Rewilding Europe, 143, 144, 153
Rhine River, 144
Rikers Island, 214
roads
    animal deaths and, 63, 64, 67
    as barriers, 14, 34, 48, 353
    increase in, 64–67
    rewilding, 330
    wildlife crossings and, 67, 70–71
Rocky Mountains, 27, 79, 176, 320
Rodriguez, Yolanda, 105
Rojas, Jose, 101
roseate spoonbills, 262, 265, 282
Rosen, Leonard and Julius, 270, 272
Rundell, Katherine, 7

S
Samburu National Reserve, 288–89, 291–92, 296–97, 306, 315
Sanderson, Eric W., 52
sandhill cranes, 71
San Francisco, 220–21, 247
Savannah River Ecology Laboratory, 56
Savannah River Site Corridor Project, 55–57, 59–61, 63
Save the Elephants (STE), 70, 184, 291, 293–94, 296, 300, 304, 306, 310, 312, 315
scarlet macaws, 109
Scarpignato, Amy, 76
Schell, Christopher, 226–27, 233–34, 238, 246–47
Schmidt, Chloé, 233–34, 242
Scott, Abby, 190
Senner, Stan, 74
Shoshone River, 176
Sibylline Mountains, 146
Sierra La Mojonera, 118, 121, 123, 135
Sierra Madre Oriental, 118
SLOSS (single large or several small), 37, 50
sloths, 83, 353
Snæfell, 335
snow geese, 2
social inequality
    biodiversity loss and, 224–30, 232–36, 238–39, 241
    wildlife observation and, 242–44, 246
solastalgia, 45
Sommers, Albert, 195
Soulé, Michael, 144
Species at Risk Act, 329
spider monkeys, 38, 107, 110
Standard Gauge Railway (SGR), 293, 308–10, 312–13, 315
Strava, 163, 165
suicide trees, 109
Swainson's hawks, 71–74
synanthropes vs. misanthropes, 203, 205

T
Tabor, Gary, 63–64
Talamanca Mountains, 82, 91–92, 98
tapirs, 90, 95

Tavone, Angela, 152–55, 160
Theobald, David, 77–79
Tibetan Plateau, 180
Toomey, Anne, 213
Topo Chico, 119
trees
    canopies, in cities, 233
    planting, 98, 100–101, 103–4
Trei, Jan-Niklas, 157
Trudeau, Justin, 329
Trump, Donald, 6, 273
Tsavo National Park, 308–10, 312, 313
turtles, 183, 279–80

U

Urban Coyote Research Program, 205
urbanization, growth of, 16–17, 52
USAID, 6
US Army Corps of Engineers, 7, 258, 261
US Atomic Energy Commission, 56
US Department of Agriculture (USDA),
    6, 191, 220, 308

V

Van Cortlandt Park, 213
van der Heiden, Craig, 264
Vasquez, Amy, 238
Vatnajökull, 335
vultures
    griffon, 148, 153, 155, 160
    important ecosystem role of, 113, 173
    king, 113, 146
    threatened, 113

W

Wallis Annenberg Wildlife Crossing, 71
Washington Square Park, 218
water. See also dams
    connectivity and, 284
    population growth and, 27, 28

    as scarce resource, 27, 119
    urban polices toward, 235–36
Weckel, Mark, 213
Whitestone Bridge, 198, 206, 213
Whitworth, Andy, 84, 86–87, 90–92, 94,
    98, 104–5, 107–11, 114
wildebeest, 35, 172, 184
The Wildlands Project, 44
wildland-urban interface (WUI), 247
Wildlife Adventures, 153
Wildlife Crossings Pilot Program, 6, 70
wildlife crossing structures, 67, 70–71
Wilkinson, Christine, 181, 183, 221, 225,
    246
Willson, Roland, 325, 329
Wilson, E. O., 35–36, 37, 38
Wittemyer, George, 184
wolverines, 199
wolves, 2, 42, 44, 142–44, 147–48, 158,
    160, 186, 191, 199–200, 325, 326
Wood, Eric, 230, 232–33, 235, 238, 239
Woodbridge, Brian, 73
wood storks, 251, 253–55, 262, 265,
    268–70, 277, 281–82, 284–85
wrentits, 71
Wyss Academy for Nature, 296

X

Xu, Wenjing, 180, 181, 189

Y

Yellowstone National Park, 44, 89, 144,
    176, 178, 190–91, 199. See also Greater
    Yellowstone Ecosystem
Yellowstone to Yukon Conservation
    Initiative (Y2Y), 42, 44

Z

zebras, 66, 143, 184, 294
zoogeochemistry, 187

Next Spread: A three-toed sloth makes its precarious way across a road on Costa Rica's Osa Peninsula. Roads serve to connect humans, but for virtually every other species on the planet, they are a barrier. *Andy Whitworth / Osa Conservation*

Back Cover: Some elephants in Kenya migrate hundreds of miles to access seasonal food sources. *Michael Nichols*